PEARSON MATH 9

Authors

Gordon Cooke

Catherine Heideman · A. J. Keene

Amy Lin · Andrew Reeves

Contributing Writers

Robert Sidley · Mary Fiore · Paul Gautreau

Duncan M. LeBlanc · Gerry Bossy

Publisher
Claire Burnett

Editorial Team
Enid Haley
Lynda Cowan
Lesley Haynes
Nirmala Nutakki
Sarah Mawson
Ioana Gagea
Cristina Getson
Lynne Gulliver
Stephanie Cox
Kaari Turk
Judy Wilson

Math Team Leader
Diane Wyman

Product Manager
Kathleen Crosbie

Editorial Contributors
Claire Sauve
May Look

Photo Research
Karen Hunter

Design
Word & Image Design Studio Inc.

Copyright © 2007 Pearson Education Canada, a division of Pearson Canada Inc.

ISBN 0-321-39313-9

Printed and bound in Canada

1 2 3 4 5 – TCP – 10 09 08 07 06

All Rights Reserved. This publication is protected by copyright, and permission should be obtained from the publisher prior to any prohibited reproduction, storage in a retrieval system, or transmission in any form or by any means, electronic, mechanical, photocopying, recording, or likewise.
For information regarding permission, write to the Permissions Department.

The information and activities presented in this book have been carefully edited and reviewed. However, the publisher shall not be held liable for any damages resulting, in whole or in part, from the reader's use of this material.

The publisher wishes to thank the staff and students of Sir Robert Borden B.T.I. for their assistance with photography.

Consultants and Advisers

Pearson Education thanks its Consultants and Advisers, who helped shape the vision for *Pearson Math 9* through discussions and reviews of prototype material and manuscript.

Assessment Consultant

Ken O'Connor

Technology Consultants

Carolyn Kieran
Tom Steinke

Literacy Consultant

Jan Crofoot

Advisers

Trevor Brown
Stewart Craven
Gord Doctorow
Dwight Stead

Reviewers

Pearson Education would like to thank the teachers and students who field-tested *Pearson Math 9* prior to publication. Their feedback and constructive recommendations have helped to develop a quality mathematics program.

Paul Alves
Peel District School Board

Yolanda Baldasaro
Niagara Catholic District School Board

Marcia Charest
Peel District School Board

Rebecca J. Cober
Thames Valley District School Board

Cathy Dunne
Peel District School Board

Roxanne Evans
Algonquin and Lakeshore Catholic District School Board

Ahmad Tariq Fahimi
Toronto District School Board

Jeff Frayne
Lambton Kent District School Board

Domenic Greto
York Catholic District School Board

Tara M. Hannah
Durham District School Board

Terry Hinan
Peel District School Board

Dianna Knight
Peel District School Board

Rodger Knight
Toronto District School Board

Peggy Leroux
Waterloo Catholic District School Board

Louis Lim
York Region District School Board

Lynn Matus
Toronto District School Board

Marsha Melnik
Toronto District School Board

Donna Moore
Peel District School Board

Charlotte Morrison
District School Board of Niagara

Priscilla Nelson
Peel District School Board

David Paddington
Lakehead District School Board

Kathy Pilon
Catholic School Board of Eastern Ontario

Lyle Robinson
Limestone District School Board

Natalie Robinson
Ottawa Carleton Catholic District School Board

Margaret Sinclair
York University

Bonnie L. Terry
Grand Erie District School Board

Karen Thomas
Catholic School Board of Eastern Ontario

Jim Vincent
Peel District School Board

Chris Wadley
Grand Erie District School Board

Table of Contents

Chapter 1: Measuring Figures and Objects

Reading and Writing in Math: Communicate Your Thinking	2
1.1 Measuring Perimeter and Area	3
1.2 Measuring Right Triangles	7
1.3 Area of a Composite Figure	11
1.4 Perimeter of a Composite Figure	15
Game: Measurement Bingo	19
Mid-Chapter Review	20
1.5 Volumes of a Prism and a Cylinder	21
1.6 Volume of a Pyramid	25
1.7 Volume of a Cone	29
1.8 Volume of a Sphere	33
Chapter Review	37
Practice Test	40

Chapter 2: Investigating Perimeter and Area of Rectangles

Reading and Writing in Math: Exploring Guided Examples	42
2.1 Varying and Fixed Measures	43
Technology: Using *The Geometer's Sketchpad* to Investigate Rectangles	47
2.2 Rectangles with Given Perimeter or Area	49
2.3 Maximum Area for a Given Perimeter	53
Technology: Using *The Geometer's Sketchpad* to Investigate Maximum Area and Minimum Perimeter	57
Game: Find the Least Perimeter	59
Mid-Chapter Review	60
2.4 Minimum Perimeter for a Given Area	61
2.5 Problems Involving Maximum or Minimum Measures	65
Chapter Review	69
Practice Test	72

v

Chapter 3: Relationships in Geometry

Reading and Writing in Math: Using a Frayer Model	74
3.1 Angles in Triangles	75
Technology: Using *The Geometer's Sketchpad* to Investigate Angles Related to Triangles	79
3.2 Exterior Angles of a Triangle	81
Technology: Using *The Geometer's Sketchpad* to Investigate Angles Involving Parallel Lines	85
3.3 Angles Involving Parallel Lines	87
Puzzle: Polygon Pieces	91
Mid-Chapter Review	92
Technology: Using *The Geometer's Sketchpad* to Investigate Angles Related to Quadrilaterals	93
3.4 Interior and Exterior Angles of Quadrilaterals	95
3.5 Interior and Exterior Angles of Polygons	99
Technology: Using *The Geometer's Sketchpad* to Investigate Angles Related to Regular Polygons	103
Chapter Review	105
Practice Test	108

Chapter 4: Proportional Reasoning

Reading and Writing in Math: Decoding Word Problems	110
4.1 Equivalent Ratios	111
4.2 Ratio and Proportion	115
Technology: Using *The Geometer's Sketchpad* to Investigate Scale Drawings	119
4.3 Unit Rates	121
Puzzle: Grid Paper Pool	125
Mid-Chapter Review	126
4.4 Applying Proportional Reasoning	127
4.5 Using Algebra to Solve a Proportion	131
4.6 Percent as a Ratio	135
Chapter Review	139
Practice Test	142
Cumulative Review Chapters 1–4	143

Chapter 5: Graphing Relations

Reading and Writing in Math: Writing Solutions	**146**
5.1 Interpreting Scatter Plots	**147**
5.2 Line of Best Fit	**151**
Technology: Using *Fathom* to Draw a Line of Best Fit	**156**
5.3 Curve of Best Fit	**159**
Technology: Using a Graphing Calculator to Draw a Curve of Best Fit	**164**
Game: Hidden Sum	**167**
Mid-Chapter Review	**168**
5.4 Graphing Linear Relations	**169**
5.5 Graphing Non-Linear Relations	**175**
5.6 Interpreting Graphs	**181**
Chapter Review	**185**
Practice Test	**188**

Chapter 6: Linear Relations

Reading and Writing in Math: Making a Mind Map	**190**
6.1 Recognizing Linear Relations	**191**
6.2 Average Speed as Rate of Change	**197**
6.3 Other Rates of Change	**201**
6.4 Direct Variation	**205**
6.5 Partial Variation	**211**
Technology: Using a Graphing Calculator to Graph an Equation and Generate a Table of Values	**217**
Game: The 25-m Sprint	**219**
Mid-Chapter Review	**220**
6.6 Changing Direct and Partial Variation Situations	**221**
Technology: Using CAS to Explore Solving Equations	**225**
6.7 Solving Equations	**227**
6.8 Determining Values in a Linear Relation	**231**
6.9 Solving Problems Involving Linear Relations	**235**
6.10 Two Linear Relations	**241**
Technology: Using a Graphing Calculator to Determine Where Two Lines Meet	**245**
Chapter Review	**247**
Practice Test	**250**

Chapter 7 — Polynomials

	Reading and Writing in Math: Doing Your Best on a Test	**252**
7.1	Like Terms and Unlike Terms	**253**
7.2	Modelling the Sum of Two Polynomials	**257**
	Technology: Using CAS to Add and Subtract Polynomials	**261**
7.3	Modelling the Difference of Two Polynomials	**263**
	Game: I Have… Who Has…?	**267**
	Mid-Chapter Review	**268**
7.4	Modelling the Product of a Polynomial and a Constant Term	**269**
	Technology: Using CAS to Investigate Multiplying Two Monomials	**273**
7.5	Modelling the Product of Two Monomials	**275**
7.6	Modelling the Product of a Polynomial and a Monomial	**279**
7.7	Solving Equations with More than One Variable Term	**283**
	Chapter Review	**287**
	Practice Test	**290**

Cumulative Review Chapters 1–7	**291**
Extended Glossary	**295**
Index	**308**
Answers	**311**
Acknowledgments	**352**

Welcome to
Pearson Math 9

This introduction will show you how to use this book. Understanding its structure can help you to be successful.

Each chapter opens with a page like this:

See **What You'll Learn** and **Why**.

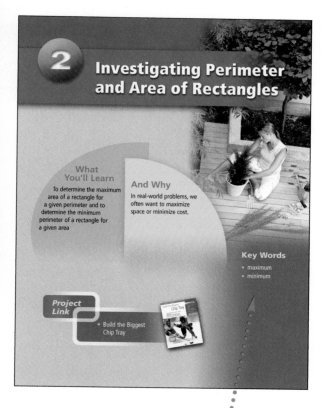

Project Link highlights a topic related to the math in the chapter.

Check the list of **Key Words**.

Reading and Writing in Math helps you develop language skills that are especially useful in learning math.

Try out some of the ideas to build your vocabulary and study skills.

In each numbered section:

Investigate a problem to develop your understanding of a new concept.

Reflect on your results with other students.

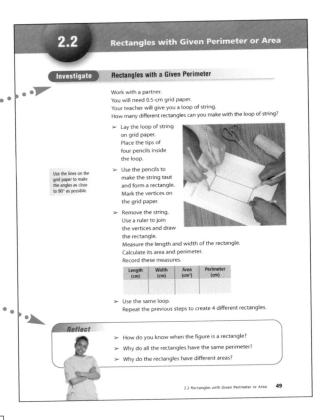

Connect the Ideas to consolidate your learning.

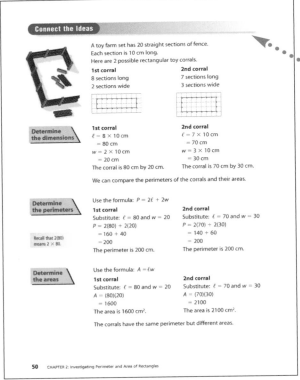

Practice questions reinforce the math.

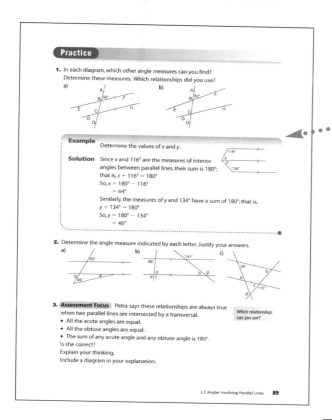

Use the guided **Example** as a model to help you solve new problems.

Try the **Take It Further** questions to extend your thinking.

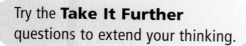

Describe your learning **In Your Own Words**.

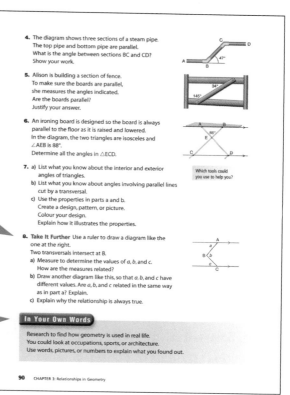

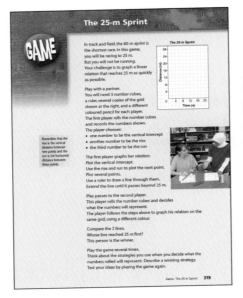

In each Chapter, play a **Game** with other students to reinforce your skills.

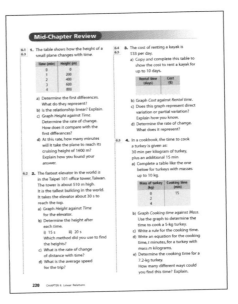

Use the **Mid-Chapter Review** to consolidate key concepts.

What Do I Need to Know?
summarizes key ideas from the Chapter.

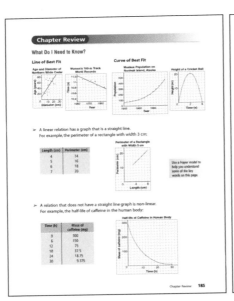

What Should I Be Able to Do? allows you to review your learning of concepts.

The **Practice Test** models the kind of questions your teacher might ask on a test.

Icons remind you that you can use **technology** tools to investigate relationships.

Follow the instructions for using **computer** software,

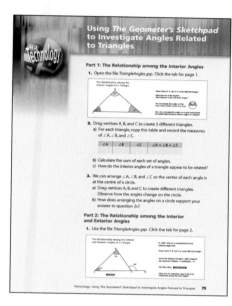
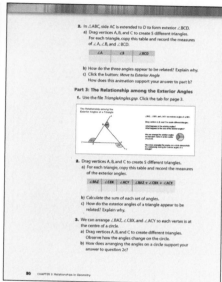

or a **calculator**.

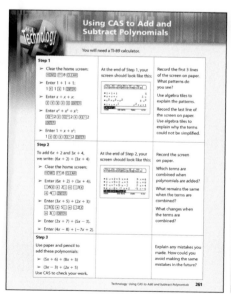
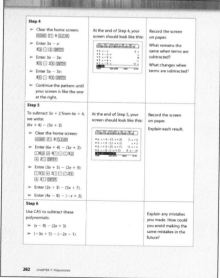

xiii

The **Extended Glossary** provides definitions for key words and concepts, and examples.

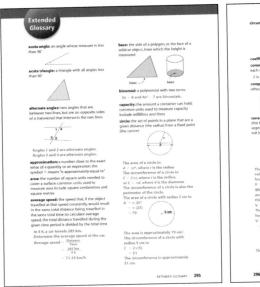
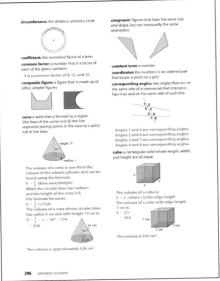

Check your **Answers** after completing a set of Practice questions.

By using the Answers responsibly, you can improve your math learning and develop good study habits.

The **CD-ROM** that accompanies this text provides:

- **Get Ready** pages for each Chapter.
  Complete the short, contextual review of content you learned in an earlier grade or chapter.

- pre-made sketches for *The Geometer's Sketchpad*

- data sets for use with *Fathom*

- copies of graphs to use as you work through the text

- an electronic version of this text

1 Measuring Figures and Objects

What You'll Learn
To measure the areas and perimeters of figures and the volumes of objects

And Why
To solve problems related to length of a fence, area of yard to be seeded, and volume of soil needed to fill a planter

Key Words
- leg
- hypotenuse
- inverse operation
- composite figure
- prism
- cylinder
- pyramid
- cone
- slant height
- sphere

Project Link

- Let's Play Mini-Golf!

Reading and Writing in Math

Communicate Your Thinking

When you solve a problem, you can communicate your thinking in words, diagrams, or numbers. Someone else should be able to understand your thinking and check your work.

Tina owns a landscape business.

Here is her sketch of a backyard and some calculations. Check Tina's calculations.

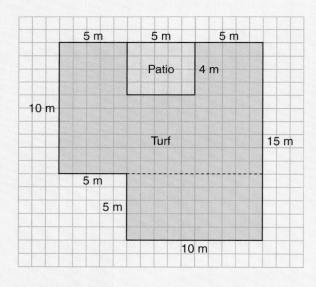

Area of turf
$5 + 5 + 5 \times 10 + 5 \times 10 - 5 \times 4$
$= 10 + 50 + 50 - 20$
$= 110 - 20$
$= 90$
I think I need 90 m^2 of turf

Write a letter to Tina.
Point out the errors she made.
Describe the errors in detail.
Provide some advice so Tina does not make the same mistakes again.
Show Tina how to communicate her thinking better.

CHAPTER 1: Measuring Figures and Objects

1.1 Measuring Perimeter and Area

Hyo Jin is redecorating her living room.
She measures the room.
Hyo Jin calculates the area of the walls to help determine how many litres of paint to buy.
She calculates the perimeter of the room to determine how many metres of wallpaper border to buy.

Investigate — Perimeter and Area Problems

Work with a partner.
You will need scissors.
Your teacher will give you a larger copy of these clues.
Cut them apart.

➤ The clues belong in sets of 4.
  Each set includes:
  1 clue with a diagram,
  1 clue describing a problem, and
  2 clues with perimeter and area formulas.
  Sort the clues into sets.

➤ Choose one set of clues.
  Solve the problem.
  Show your work.

Reflect

➤ How did you solve the problem you selected?

➤ What is another way to solve the same problem?

Connect the Ideas

The perimeter of a figure is the distance around it.

The area of a figure is the number of square units inside it.

One way to calculate perimeter or area is to substitute the appropriate measures into a formula.

Rectangle

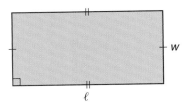

$P = 2\ell + 2w$
$A = \ell w$

Square

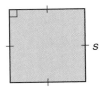

$P = 4s$
$A = s^2$

You can also find the perimeter of a figure whose sides are line segments by adding the lengths of its sides. You do not have to memorize the perimeter formulas.

Parallelogram

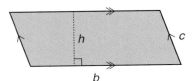

$P = 2b + 2c$
$A = bh$

Triangle

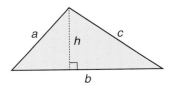

$P = a + b + c$
$A = \frac{1}{2}bh$

The circumference of the circle is also the perimeter of the circle.

Circle

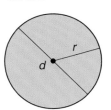

$C = \pi d$ or $C = 2\pi r$
$A = \pi r^2$

Trapezoid

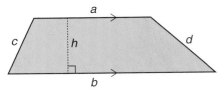

$P = a + b + c + d$
Area = $\frac{1}{2}$ (sum of parallel sides) × height
or, $A = \frac{1}{2}(a + b)h$

Practice

1. Determine the perimeter and area of each figure.

a) 9 m, 5 m (rectangle)

b) circle, radius 8 cm

c) right triangle, 10 cm, 24 cm, 26 cm

2. Determine the perimeter and area of each figure.

a) parallelogram, 8 cm base, 5 cm side, 3 cm height

b) trapezoid, 10 cm top, 24 cm bottom, 15 cm and 13 cm sides, 12 cm height

c) semicircle, radius 5 cm

3. Determine the perimeter and area of each figure.

a) rhombus, 3.7 cm side, 2.8 cm height

b) triangle, 2.5 cm, 3.5 cm sides, 5.2 cm base, 1.5 cm height

c) trapezoid, 14.0 cm top, 9.0 cm bottom, 4.9 cm and 5.4 cm sides, 4.5 cm height

4. The sail on a yacht has the shape of a triangle. What is the area of this sail?

2.1 m, 2.5 m sides, 3.5 m base, 1.5 m height

Remember to show your work.

5. Reanne is making the circles of the Olympic symbol from plastic tubing. Each circle has radius 75 cm. How much tubing does she need?

6. When a paper towel tube is cut along its seam, it unwraps to form a parallelogram. How much cardboard is used to make the tube?

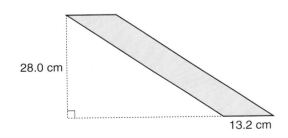

28.0 cm, 13.2 cm

1.1 Measuring Perimeter and Area   **5**

We can use a formula to determine the length or width when the perimeter is known.

> **Example**
>
> The perimeter of a rectangle is 56 cm. Its width is 4 cm. What is its length?
>
> **Solution**
>
> The perimeter P of a rectangle is:
> $P = 2\ell + 2w$
> Substitute: $P = 56$ and $w = 4$
> $56 = 2\ell + 2(4)$
> Solve for ℓ.
> $56 = 2\ell + 8$ Think: What do we add to 8 to get 56?
> We know that $56 = 48 + 8$
> So, $2\ell = 48$ Think: What do we multiply 2 by to get 48?
> We know that $2 \times 24 = 48$
> So, $\ell = 24$
> The rectangle is 24 cm long.

7. The area of a rectangle is 48 cm².
   a) The width is 6 cm. What is its length?
   b) The length is 12 cm. What is its width?

Having trouble? Read the Example above.

8. Rosa has 24 m of fencing to make a square pen for her dog. How long is each side of the pen? Sketch the pen. Justify your answer.

9. **Assessment Focus** Serena has 3 m of garden edging. She wants to make a flowerbed that is an isosceles triangle.
   a) Suppose each equal side is 90 cm long. How long is the third side?
   b) Suppose the third side is 90 cm long. How long is each equal side?
   Justify your answers.

Recall that 1 m = 100 cm.

10. **Take It Further** Luis makes a circle from a piece of wire 120 cm long.
    a) What is the diameter of the circle?
    b) Will the wire fit around a circular tube with diameter 40 cm? Justify your answer.

In Your Own Words

Choose one of the figures from this section.
Explain how you found its perimeter or area.

1.2 Measuring Right Triangles

People in many occupations, such as carpenters, plumbers, and building contractors, use right triangles in their work.

Investigate — Squares Drawn on the Sides of Triangles

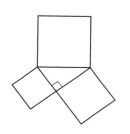

You will need a ruler and protractor.
Your teacher will give you a larger copy of these triangles.

Each triangle has a square drawn on each side.
The side length of the square is the side length of the triangle.

➤ Begin with the right triangle.
  Calculate the area of each square.
  What do you notice about the 3 areas?

➤ Use a ruler and protractor.
  Draw a different right triangle and the squares on its sides.
  Are the areas of the squares related in the same way?

➤ Do you think the areas are related in the same way for each of these triangles?

  • an acute triangle   • an obtuse triangle

  To find out, repeat the activity for these triangles.

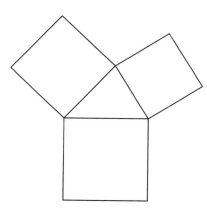

Reflect

➤ Work with a classmate.
  Write a word equation that relates the areas of the squares on the sides of a right triangle.
  Include a sketch of a right triangle.

➤ Are the areas of the squares on the sides of an acute or obtuse triangle related in the same way?
  Justify your answer.

Connect the Ideas

In a right triangle, the shorter sides are the **legs**. These sides form the right angle. The longest side is opposite the right angle. It is called the **hypotenuse**.

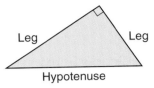

The sum of the areas of the squares on the legs is equal to the area of the square on the hypotenuse.

This relationship is written as:
$a^2 + b^2 = c^2$

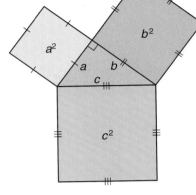

This relationship is named the Pythagorean Theorem. The theorem is not true for acute or obtuse triangles. This relationship can be used to determine the length of the hypotenuse of a right triangle when you know the lengths of the legs.

To determine the hypotenuse in this right triangle, substitute for a and b in the formula $a^2 + b^2 = c^2$.
Substitute: $a = 7$ and $b = 5$

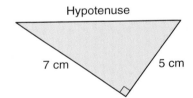

$$7^2 + 5^2 = c^2$$
$$(7 \times 7) + (5 \times 5) = c^2$$
$$49 + 25 = c^2$$
$$74 = c^2 \quad \text{Use the \textbf{inverse operation}.}$$
$$c = \sqrt{74} \quad \text{You know } c^2. \text{ To find } c, \text{ take the square root.}$$
$$c \doteq 8.602 \quad \text{Use the } \sqrt{} \text{ key on a calculator.}$$

The hypotenuse is about 8.6 cm, to 1 decimal place.
This result is reasonable. That is, the hypotenuse is always greater than the legs and 8.6 is greater than 5 and 7.

Practice

Where necessary, give the answers to 1 decimal place.

1. Determine each unknown length.

   a) [triangle with 8 cm, 6 cm, right angle]

   b) [triangle with 5 cm, 12 cm, right angle]

   c) 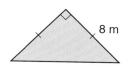 [triangle with 8 m, right angle]

2. Determine each unknown length.

   a) [triangle with 3.4 cm, 5.9 cm, right angle]

   b)  [triangle with 4.6 m, right angle]

   c) 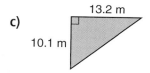 [triangle with 13.2 m, 10.1 m, right angle]

3. Ali walks along the path through the park. How far does Ali walk?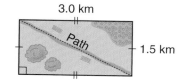

4. A right triangle has legs 9 cm and 12 cm. Sketch the triangle. What is its perimeter? Show your work.

5. A plumber is installing pipe. She has to offset the pipe around an obstacle.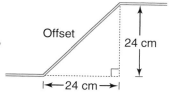
   a) What is the length of this offset section of pipe? Justify your answer.
   b) Is the result reasonable? Explain.

6. A contractor estimates how many sheets of plywood will be needed to build this roof. His first step is to determine the length of the sloping part. How long is it? Justify your answer.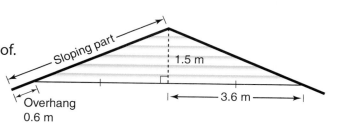

7. Use the Pythagorean Theorem to find out if these are right triangles. Justify your answers.

   a) [triangle with 3 cm, 4 cm, 5 cm]

   b) 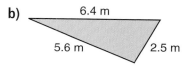 [triangle with 6.4 m, 5.6 m, 2.5 m]

   c) 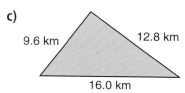 [triangle with 9.6 km, 12.8 km, 16.0 km]

Remember: $a^2 + b^2 = c^2$ is only true for right triangles.
c is the longest side.

We can also use the Pythagorean Theorem to determine the length of a leg.

Example

Kim is building a ramp with a piece of wood 175 cm long.
The height of the ramp is 35 cm.
What is the horizontal length of the ramp?

Solution

Since the side view of the ramp is a right triangle, we use the Pythagorean Theorem.
Let the horizontal leg be b.

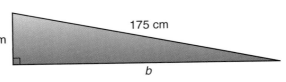

Use: $a^2 + b^2 = c^2$
Substitute: $a = 35$ and $c = 175$
$$35^2 + b^2 = 175^2$$
$$1225 + b^2 = 30\,625$$
$$1225 - 1225 + b^2 = 30\,625 - 1225$$
$$b^2 = 29\,400$$
$$b = \sqrt{29\,400}$$
$$b \doteq 171.46$$

To isolate b^2, subtract 1225 from each side of the equation.
To calculate b, take the square root.

The horizontal length of the ramp is about 171 cm.

8. Determine each unknown length.

a)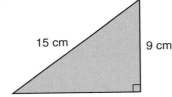

b)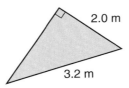

c)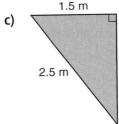

9. **Assessment Focus** Suppose you know that the lengths of two sides of a right triangle are 3.5 cm and 4.5 cm. What is the length of the third side? Show two possible answers.

10. **Take It Further** A ladder is 4.9 m long. It leans against a wall with its foot 1.2 m from the base of the wall. The distance from the foot of a ladder to the wall should be about one-quarter of the distance the ladder reaches up the wall.
Is the ladder safely positioned? Justify your answer.

In Your Own Words

Describe how the areas of the squares drawn on the sides of a right triangle are related. Why is it important to know the Pythagorean Theorem?

1.3 Area of a Composite Figure

An appraiser is determining the value of some vacant land.
As part of the process, he must calculate the area of the land.
Often parcels of land have irregular shapes.
How can their areas be calculated?

Investigate — Drawing a Composite Figure and Determining Its Area

You may need dot paper or grid paper.
You know how to determine the area of each figure below.

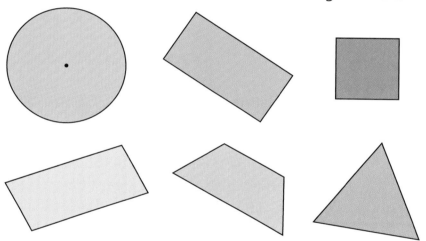

- Combine any or all of these figures, or parts of them.
  Design an unusual garden.

- Determine the area of your garden.
  Write enough measures on your design so someone else could calculate its area.

Reflect

- Trade designs with a classmate.
  Calculate the area of your classmate's garden.

- Compare answers. If both of you have different answers for the same garden, try to find out why.

Keep your designs for use in Section 1.4.

Connect the Ideas

A figure made up of other figures is called a **composite figure**. This composite figure is made up of a rectangle and a trapezoid.

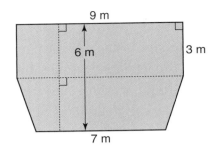

Determine the area of each part

The rectangle has dimensions 9 m by 3 m.
Its area is: $A = \ell w$
$= 9 \times 3$
$= 27$

The trapezoid has parallel sides of 9 m and 7 m.
The height of the trapezoid is: $6 \text{ m} - 3 \text{ m} = 3 \text{ m}$
Its area is: $A = \frac{1}{2}(a + b)h$
$= \frac{1}{2}(9 + 7) \times 3$
$= \frac{1}{2} \times 16 \times 3$
$= 8 \times 3$
$= 24$

Determine the total area

Total area = Rectangle area + Trapezoid area
$= 27 + 24$
$= 51$

The area of the composite figure is 51 m².

Is the answer reasonable?

From the diagram, the area of the trapezoid is a little less than the area of the rectangle. So, the total area should be less than twice the area of the rectangle. That is, less than $2 \times 27 = 54$

Since 51 is a little less than 54, the answer is reasonable.

Practice

Assume the figures in questions 1 and 2 are drawn on 1-cm grid paper.

1. Determine the area of each composite figure.

   a)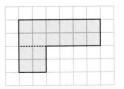

   b)

   c)

12   CHAPTER 1: Measuring Figures and Objects

2. Determine the area of each composite figure.
   a)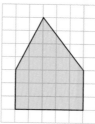
   b)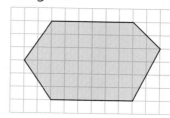
   c)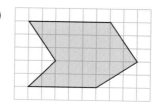

3. Determine the area of each composite figure.
   a)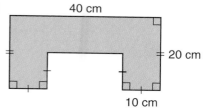
   b)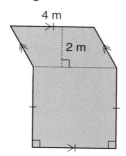
   c)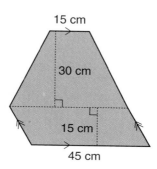

Another way to determine the area of a composite figure is to draw a figure around the composite figure first.

Example

Determine the area of this composite figure. The curve is a semicircle.

Solution

The composite figure is a rectangle that measures 9.2 cm by 13.0 cm, with a semicircle removed.

- The diameter of the semicircle is 9.2 cm.
  So, the radius of the semicircle is:
  $\frac{9.2 \text{ cm}}{2} = 4.6 \text{ cm}$
  The area of a circle is: $A = \pi r^2$
  So, the area of the semicircle is:
  $A = \frac{1}{2}\pi r^2$
  Substitute: $r = 4.6$
  $A = \frac{1}{2} \times \pi \times (4.6)^2 \doteq 33.24$
- The area of the rectangle is: $9.2 \times 13.0 = 119.6$
- Total area = Rectangle area − Semicircle area
  $\doteq 119.6 - 33.24$
  $= 86.36$

The area of the composite figure is about 86.4 cm².

1.3 Area of a Composite Figure

4. Calculate the area of each figure.
   a)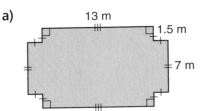
   b)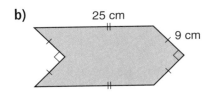

5. Kiren quotes for paving driveways based on the area to be paved. Determine the area of each driveway. All curves are semicircles. Is each result reasonable? Explain.
   a)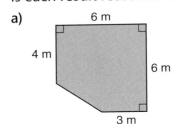
   b)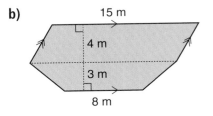
   c)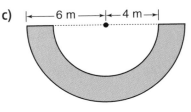

6. **Assessment Focus** The floor plan of a sunroom is shown.
   a) What is the area of the floor of this sunroom?
   b) There are 10 000 cm² in 1 m².
      What is the area of the floor in square centimetres?
   c) A square tile has side length 30 cm.
      What is the area of 1 tile?
   d) Estimate how many tiles are needed to cover the floor. Justify your answer.

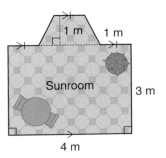

7. **Take It Further** One flowerbed has the shape of a trapezoid.
   The parallel sides are 5 m and 7 m long.
   The distance between the parallel sides is 4 m.
   On the shorter parallel side, there is another flowerbed that has the shape of a rhombus, with side length 5 m and height 3 m.
   a) Sketch the flowerbeds.
   b) What is the area of the flowerbeds?
   c) One bag of topsoil covers 0.25 m². How many bags are needed?
   d) One bag of topsoil costs $2.29, including taxes.
      How much will the topsoil in part c cost?

In Your Own Words

Sketch and label a composite figure.
Determine its area. Try to do this two different ways if possible.
Show your work.

1.4 Perimeter of a Composite Figure

Tamara works for a fencing company. She is preparing a price quote for a customer. Tamara needs to know the type of fencing being ordered and the perimeter of the area to be fenced.

Investigate — Perimeter of a Composite Figure

Your teacher will give you a large copy of this design. The curve is a semicircle.

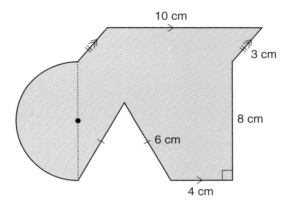

Or, you can use your garden design from Section 1.3. Calculate the perimeter of the design.

Reflect

➢ Trade solutions with a classmate. Check your classmate's solution.

➢ Compare answers. If you have different answers for the same garden, try to find out why.

➢ For which figures are the measurements you need for perimeter different from those you need for area? How did you find these measurements?

Connect the Ideas

Determine the curved length

Here is a composite figure from Section 1.3.
The perimeter of this figure is the sum of
3 sides of a rectangle and
one-half the circumference of a circle.
The diameter of the circle is 9.2 cm.
The circumference of a circle is: $C = \pi d$
So, the circumference is: $C = \pi \times 9.2$
One-half the circumference is: $\frac{1}{2} \times \pi \times 9.2 \doteq 14.45$

Determine the perimeter

The approximate perimeter of the composite figure is:
$13.0 + 9.2 + 13.0 + 14.45 = 49.65$
The perimeter is about 49.7 cm.

Check the result

From the diagram, the length of the semicircle
is greater than the width of the rectangle.
So, the perimeter of the figure should be greater than
the perimeter of the rectangle, which is approximately
$2(9) + 2(13) = 18 + 26 = 44$; the result is reasonable.

Practice

1. a) Determine the perimeter of each composite figure.

i) 6.0 cm, 4.0 cm

ii) 6.0 cm, 4.0 cm

iii) 6.0 cm, 4.0 cm

iv) 6.0 cm, 4.0 cm

b) What do you notice about the perimeters in part a?
Do you think the same relationships are true
for the areas? How could you find out?

2. Sarah is lighting the theatre stage arch for the new play.
The arch is a semicircle on top of a rectangle.
How long is the string of lights? Justify your answer.

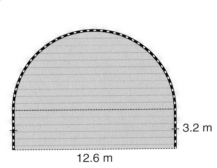
3.2 m
12.6 m

16 CHAPTER 1: Measuring Figures and Objects

3. This design is 4 one-quarter circles on the sides of a square.
   a) What is the perimeter of the design? The broken lines are not part of the perimeter.
   b) Is your result reasonable? Explain.

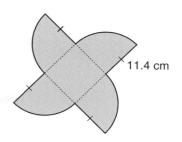

Sometimes you may need to use the Pythagorean Theorem to calculate a length before you can determine the perimeter.

Example

Here is a plan of a driveway from Section 1.3.

A fence is to be placed around the driveway on the sides indicated. How much fencing is needed?

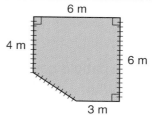

Solution

We know the length of each part of the fence except for AB. Draw right △ABC.
Then AC = 6 m − 4 m = 2 m
and BC = 6 m − 3 m = 3 m
To find the length of AB, use the Pythagorean Theorem in △ABC.

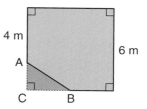

$a^2 + b^2 = c^2$

Substitute: $a = 3$ and $b = 2$
$3^2 + 2^2 = c^2$
$9 + 4 = c^2$
$13 = c^2$
$c = \sqrt{13}$
$c \doteq 3.6$

The total length of fencing is:
6 m + 3.6 m + 4 m = 13.6 m
About 14 m of fencing are needed.

4. Determine the perimeter of each figure. Show your work.

   a) 23.0 cm, 11.0 cm, 5.0 cm

   b) 12.0 cm, 10.0 cm, 12.0 cm

   Need Help?
   Read the Example on page 17.

5. **Assessment Focus** A circular fish pond is set in a rectangular patio.

   a) Plastic edging is placed around the pond and the patio.
      i) What length of edging is used?
      ii) The edging costs $4.79/m.
         What is the total cost of the edging?
   b) The patio is paved with sandstone tiles.
      i) What is the area that is paved?
      ii) The sandstone costs $45.00/m^2.
         What is the total cost of the sandstone?
   c) What assumptions did you make in parts a and b?

6. Sketch two different composite figures that have the same perimeter. Calculate the perimeters, or explain how you know they are equal. Calculate each area.

7. **Take It Further** Yazan is putting up a wallpaper border in his family room. The border will run along the top of all the walls, including above any doors and windows.

   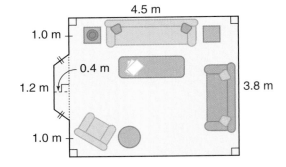

   a) What length of border does Yazan need?
   b) The border comes in 4.57-m rolls. How many rolls does Yazan need?
   c) Each roll of border costs $14.99. How much will the border for this room cost?

In Your Own Words

Sketch a composite figure.
Explain how you calculate its perimeter.
Show your work.

Measurement Bingo

Materials:
- counters as markers
- 2-cm grid paper
- scissors
- container

Play in a group of 4.

➤ Each player outlines a 4 by 4 square on 2-cm grid paper.

➤ Write these 16 answers anywhere on your 4 by 4 grid — one in each square:
circumference of a circle; area of a circle; area of a square; trapezoid; cm^3; m^2; 6; 9; 10; 11; 20; 24; 27; 30; 34; 45

➤ Your teacher will give your group a copy of 16 questions.

➤ Cut the questions apart.
Fold each question, and place it in a container.

➤ Take turns to pick a question.
Everyone answers the question.
Cover the answer on your grid with a marker.

➤ The first person to cover a row or column wins.

Mid-Chapter Review

1.1 **1.** For each figure:
- Determine its area.
- Determine its perimeter.

a) b)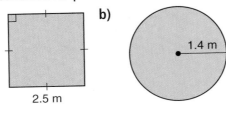

c)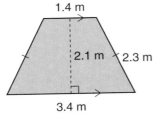

2. For each measurement in question 1, did you use a formula? If your answer is yes, write the formula. If your answer is no, explain how you calculated.

1.1
1.2 **3.** Determine the area and perimeter of each figure.

a) 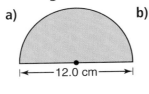 b)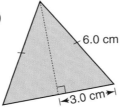

1.2 **4.** Determine each unknown length.

a)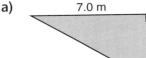

b)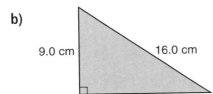

5. Determine the perimeter of each triangle.

a) 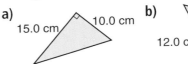 b)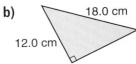

6. Determine each unknown length.

a) b)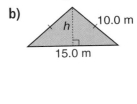

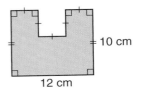

1.3
1.4 **7.** Determine the perimeter and area of each composite figure. How do you know your results are reasonable?

a) A rectangle with a square removed

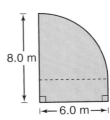

b) A quarter circle on a rectangle

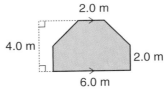

1.3 **8.** Andrew is painting 2 coats on 2 walls of his studio. Each wall looks like this:

One can of paint covers 10.5 m² with one coat. How many cans of paint does Andrew need? Justify your answer.

20 CHAPTER 1: Measuring Figures and Objects

1.5 Volumes of a Prism and a Cylinder

Some buildings and bridges are supported by concrete columns called piers. A civil engineer calculates the volume of the piers so that enough concrete is ordered for a project.

Investigate: Relating the Volumes of a Prism and a Cylinder

Which of these pictures represent prisms? Justify your answers. What would you need to know to determine the volume of each prism?

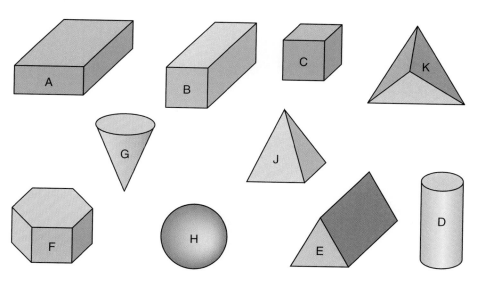

Reflect

> What is the same about all the prisms? What is different?

> Suppose each prism is filled with layers of congruent figures. Which figure would it be in each prism?

> Which object is a cylinder? Choose one of the prisms and compare it with the cylinder. What is the same about these two objects? What is different?

1.5 Volumes of a Prism and a Cylinder **21**

Connect the Ideas

Compare a prism and a cylinder.
Each solid can be placed with its top directly above the base.
Visualize slicing the prism and the cylinder into layers.

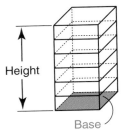 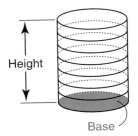

The area of each layer equals the area of the base.
The height of the layers is equal to the height of the prism and cylinder.
So, the volume of the solid equals the area of the base multiplied by the height.

The volume of a prism is:	The volume of a cylinder is:
V = base area $\times$ height	V = base area $\times$ height

The base area of a cylinder is: πr^2
The height is: h

So, the volume of a cylinder can also be written as:
$V = \pi r^2 h$

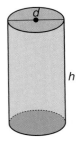

A hot-water tank is shaped like a cylinder with base diameter 56 cm and height 120 cm.
Its volume can be found using this formula: $V = \pi r^2 h$
The radius r is: $\frac{56 \text{ cm}}{2} = 28$ cm

Substitute: $r = 28$ and $h = 120$
$V = \pi \times 28^2 \times 120$
$\doteq 295\,561.0$

The volume of the tank is about 300 000 cm³.
1 cm³ of volume = 1 mL of capacity
And, 1000 mL = 1 L
So, 300 000 cm³ = $\frac{300\,000}{1000}$ L
       = 300.000 L
The capacity of the tank is about 300 L.

A label on the tank indicates that its capacity is 190 L.

Why do you think the two capacities are so different?

Practice

1. Determine the volume of each prism.
   a)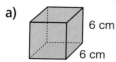
   b)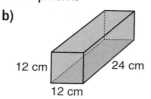
   c)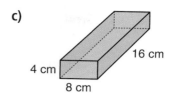

2. Determine the volume of each prism.
   a)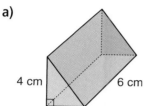
   b)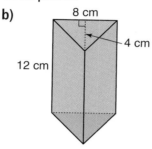
   c)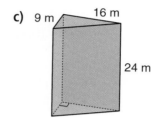

   Need Help?
   Read Connect the Ideas.

3. Determine the volume of each cylinder.
   a)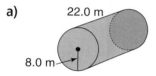
   b)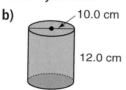
   c)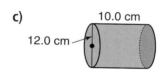

4. Pasta is sold in a box that is a rectangular prism. The box that feeds 4 people measures 3 cm by 9 cm by 18 cm.
   a) What is the volume of this box?
   b) The company wants to produce a party-pack box of pasta. Each dimension of the box will be doubled. Will this be enough pasta for 16 people? Justify your answer.

   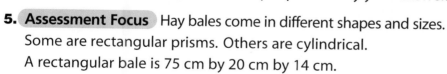

5. **Assessment Focus** Hay bales come in different shapes and sizes. Some are rectangular prisms. Others are cylindrical.
   A rectangular bale is 75 cm by 20 cm by 14 cm.
   A cylindrical bale has base diameter 150 cm and length 120 cm.
   a) Sketch each bale. Which has the greater volume? Justify your answer.
   b) About how many of the smaller bales have a total volume equal to that of the larger bale?

6. a) What is the volume of this barn?
   b) Would this barn hold 1000 of the rectangular hay bales in question 5? How do you know?
   c) Would this barn hold 1000 of the cylindrical hay bales in question 5? How do you know?
   d) Are your results reasonable? Explain.

1.5 Volumes of a Prism and a Cylinder

Sometimes we need to use the Pythagorean Theorem to calculate a length on a prism, before we find its volume.

Example

a) Determine the height of the base of this prism.
b) Determine the volume of this prism.

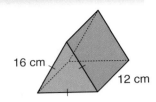

Solution

A base of a prism is not necessarily the bottom face.

a) Sketch the triangular base.
Let the height of the triangle be h.
The height bisects the base of the triangle.
Use the Pythagorean Theorem in $\triangle ABC$.

$h^2 + 8^2 = 16^2$
$h^2 + 64 = 256$
$h^2 = 256 - 64$
$h^2 = 192$
$h = \sqrt{192}$
$h \doteq 13.86$

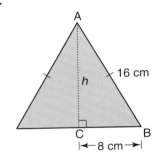

The height of the base is about 14 cm.

b) The base area is: $\frac{1}{2} \times 16 \times 13.86 = 110.88$
The length is: 12 cm
The volume is: base area $\times$ length $= 110.88 \times 12$
$= 1330.56$

The volume of the prism is about 1331 cm³.

7. A child's building block set has these triangular prisms. Determine the volume of wood in each block.

   a) Equilateral triangular prism

   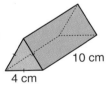

   b) Isosceles triangular prism

   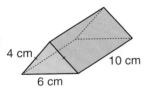

8. **Take It Further** Many types of cheese are produced in cylindrical slabs. One-quarter of this slab has been sold.
   a) What is the volume of this piece of cheese?
   b) The mass of 1 cm³ of cheese is about 1.2 g. What is the mass of the cheese shown here?

   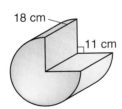

In Your Own Words

Why can you use the same formula to calculate the volumes of a prism and a cylinder? Include examples in your explanation.

1.6 Volume of a Pyramid

A company wants to know how much oil is contained in a new oilfield. A geologist estimates the volume of oil underground. She models the layers of oil as cut-off pyramids and calculates their volumes.

Investigate — Relating the Volumes of a Prism and a Pyramid

A prism and pyramid with the same base and height are related.

Work with a partner.
You will need:
- a pyramid and a prism with congruent bases and equal heights
- sand or plastic rice

➢ Make a prediction. How do you think the volumes of this pyramid and prism compare?

Fill the pyramid with sand.

➢ Estimate how many pyramids of sand will fill the prism.

➢ Fill the prism to check your estimate.

➢ How is the volume of the pyramid related to the volume of the prism? How does this compare with your prediction?

Reflect

Compare results with classmates who used related prisms and pyramids with bases different from yours.

➢ Does the volume relationship depend on the shape of the base of the related pyramid and prism? Explain.

➢ Use what you know about the volume of a prism to write a formula for the volume of a related pyramid.

Connect the Ideas

The contents of three pyramids fit exactly into the prism.
These 3 volumes together…   …are equal to this volume.

That is, the volumes of 3 pyramids are equal to the volume of the related prism.
So, the volume of a pyramid is one-third the volume of the related prism.

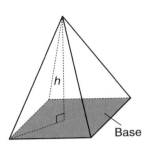

$V = \frac{1}{3}Bh$, where B is the area of the base of the pyramid and h is the height of the pyramid.

To calculate how much plaster is needed to fill this mould, we calculate the volume of the pyramid.
The base of the pyramid is a square with side length 22 cm.
So, the base area is: $B = 22 \times 22 = 484$
The height of the pyramid is: $h = 65$
The volume of the pyramid is: $V = \frac{1}{3}Bh$

$$V = \frac{1}{3} \times 484 \times 65$$
$$\doteq 10\,486.67$$

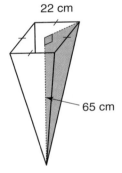

The volume of plaster is about 10 487 cm³.

Practice

1. The prism and pyramid have the same base and height.
   Determine each volume.

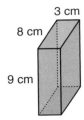

26   CHAPTER 1: Measuring Figures and Objects

2. Determine the volume of each rectangular pyramid.

a)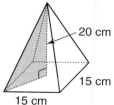
b)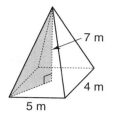
c)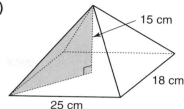

3. Determine the volume of each pyramid.

a)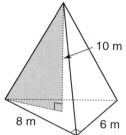
b)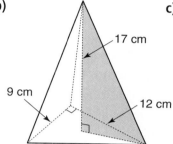
c)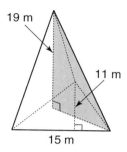

4. Pyramids have been constructed in many places around the world. One of the most famous is the Great Pyramid of Giza. It contains the burial chamber of Pharaoh Khufu. Today, the pyramid is 137 m high. When first constructed, it was 146.5 m high.
  a) Sketch the original pyramid. Label the sketch with the given measurements.
  b) What volume of rock has been lost over the years? Why do you think it has been lost?

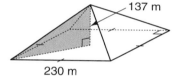

5. A pyramid in Pune, India, can hold 5000 people.
  a) What is the volume of air per person in the pyramid?
  b) How do you know your answer is reasonable?

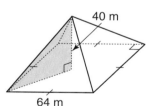

6. **Assessment Focus** A package for a frozen treat is a triangular pyramid. All edge lengths are 12.0 cm. Each triangular face has height 10.4 cm.
  a) Calculate the volume of the pyramid. Show your work.
  b) The package lists its contents as 200 mL. Why are the contents in millilitres different from the volume in cubic centimetres?

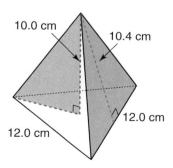

$1\ cm^3 = 1\ mL$

1.6 Volume of a Pyramid

Sometimes the height of a pyramid is difficult to measure.
We can use the Pythagorean Theorem to calculate the height of the pyramid.

Example

A crystal paperweight is a pyramid with dimensions as shown.
a) What is the height of the pyramid?
b) What is the volume of crystal in the pyramid?

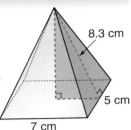

Solution

a) The height of the pyramid is one leg of a right triangle. The hypotenuse is 8.3 cm. The other leg is one-half the base of the other triangular face, or 3.5 cm. Use the Pythagorean Theorem in △ABC.

$h^2 + 3.5^2 = 8.3^2$
$h^2 + 12.25 = 68.89$
$h^2 = 68.89 - 12.25$
$h^2 = 56.64$
$h = \sqrt{56.64}$
$h \doteq 7.526$

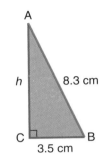

The height of the pyramid is about 7.5 cm.

b) The volume of the pyramid is: $V = \frac{1}{3}Bh$
The base area is: $B = 7 \times 5 = 35$
So, $V = \frac{1}{3} \times 35 \times 7.526 \doteq 87.80$
The volume of crystal in the pyramid is about 88 cm³.

7. A wooden pyramid has a square base with side length 14.0 cm. The height of each triangular face is 16.8 cm.
   a) Determine the height of the pyramid.
   b) Determine the volume of wood in the pyramid.

8. **Take It Further** The volume of the Great Pyramid of Cholula in Mexico is estimated to be 4.5 million cubic metres. It is 66 m high. What is the side length of its square base? Justify your answer.

In Your Own Words

How is the volume of a pyramid related to the volume of a prism? Include diagrams in your explanation.

1.7 Volume of a Cone

Large quantities of grains are sometimes stored in conical piles.
Often, only the height of the pile is known. Techniques have been developed to estimate the radius of the pile.
The volume can then be calculated.

Investigate — Relating the Volumes of a Cylinder and a Cone

A cylinder and cone with the same base and height are related.

Work with a partner.
You will need:
- a cylinder and a cone with congruent bases and equal heights
- sand or plastic rice

➤ Make a prediction. How do you think the volumes of this cylinder and cone compare?

Fill the cone with sand.

➤ Estimate how many cones of sand will fill the cylinder.

➤ Fill the cylinder to check your estimate.

➤ How is the volume of the cone related to the volume of the cylinder? How does this compare with your prediction?

Reflect

Compare results with classmates who used related cylinders and cones with bases different from yours.

➤ Does the volume relationship depend on the size of the base of the related cone and cylinder? Explain.

➤ Use what you know about the volume of a cylinder to write a formula for the volume of a related cone.

Connect the Ideas

A cone and a cylinder with the same base and height are related.

The relationship between the volumes of a cone and its related cylinder is the same as that for a pyramid and its related prism.

These 3 volumes together …   … are equal to this volume.

The volume of a cone is one-third the volume of its related cylinder.

So, the volume of a cone is: $V = \frac{1}{3}Bh$

where B is the area of the base of the cone and h is the height of the cone.

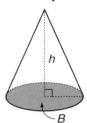

The volume of a cylinder is: $V = \pi r^2 h$
So, the volume of its related cone is:
$V = \frac{1}{3}\pi r^2 h$ or $\frac{\pi r^2 h}{3}$ or $\pi r^2 h \div 3$

The volume of a cone is $V = \frac{1}{3}\pi r^2 h$, where r is the base radius and h is the height.

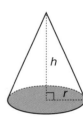

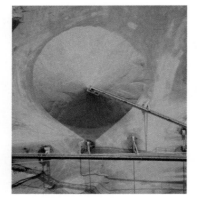

When a conveyor belt drops gravel, the gravel forms a cone.
This cone is 3.7 m high and has a base diameter of 4.6 m.

The volume of gravel is: $V = \frac{1}{3}\pi r^2 h$
$r = \frac{4.6 \text{ m}}{2} = 2.3$ m
Substitute: $r = 2.3$ and $h = 3.7$
$V = \frac{1}{3} \times \pi \times 2.3^2 \times 3.7$
$V \doteq 20.497$
The volume of gravel is about 20.5 m³.

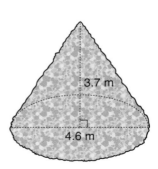

30   CHAPTER 1: Measuring Figures and Objects

Practice

1. The cylinder and cone in each pair have the same base and height. The volume of each cylinder is given. Determine the volume of each cone.

a)
$V = 42$ cm³ $V = ?$

b)
$V = 19.2$ cm³ $V = ?$

Need Help?
Read Connect the Ideas.

2. The cylinder and cone in each pair have the same base and height. Determine each volume.

a)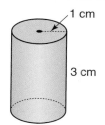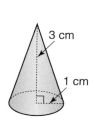

b)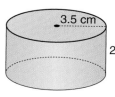

3. Determine the volume of each cone.

a)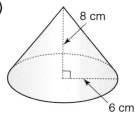

b)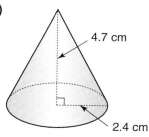

c)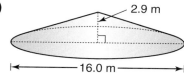

4. An ice-cream cone has diameter 6 cm and height 12 cm. What is the volume of the cone? Justify your answer.

Sometimes the height of a cone is not given. We can measure the **slant height** and radius, then use the Pythagorean Theorem to calculate the height.

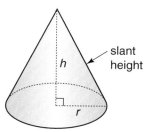

Example

A paper drinking cup is a cone.
The base has diameter 6.4 cm.
The slant height is 9.5 cm.
a) Determine the height of the cone.
b) Determine the volume of water that will fill the cup.

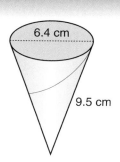

1.7 Volume of a Cone   **31**

Solution

a) The radius of the cone is: $r = \frac{6.4 \text{ cm}}{2} = 3.2$ cm

Use the Pythagorean Theorem:
$$h^2 + 3.2^2 = 9.5^2$$
$$h^2 + 10.24 = 90.25$$
$$h^2 = 90.25 - 10.24$$
$$h^2 = 80.01$$
$$h = \sqrt{80.01}$$
$$h \doteq 8.945$$

The cone is about 9 cm high.

b) The volume of water the cone will hold is equal to the volume of the cone.

Use: $V = \frac{1}{3}\pi r^2 h$

Substitute: $r = 3.2$ and $h = 8.945$

$V = \frac{1}{3} \times \pi \times 3.2^2 \times 8.945$

$V \doteq 95.92$

The volume of water is about 96 cm³.

---

5. Determine the height and volume of each cone.

a)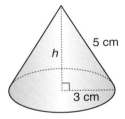

b)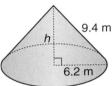

c)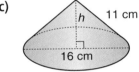

6. Assessment Focus A pile of sand is a cone.
The base diameter is 3.8 m and the slant height is 4.0 m.
a) What is the volume of sand in the pile? Include a diagram.
b) How do you know your answer is reasonable?

7. Take It Further A hill can be approximated as a cone.
Its circumference is about 3 km. Its slant height is about 800 m.
What is the approximate volume of soil in the hill?

Recall that
1 km = 1000 m.

In Your Own Words

How is the volume of a cone related to the volume of a cylinder?
Include diagrams in your explanation.

1.8 Volume of a Sphere

Some companies have experimented with making ball bearings in space. In space, a molten lump of steel floats. As the steel cools and hardens, it forms a perfect sphere. This is simpler than the manufacturing process on Earth. How could an engineer determine how much molten steel would be needed to produce a particular size of sphere in space?

Investigate — The Volume of a Sphere

Recall that
1 mL = 1 cm³.

Work with a partner.
You will need:
- a graduated cylinder
- a sphere that sinks
- water

➤ Estimate the volume of the sphere. Explain your estimation strategy.

➤ Record the volume of water in the graduated cylinder.
Place the sphere in the cylinder.
Record the new level of water.
Calculate the volume of the sphere in cubic centimetres.
How does the actual volume compare with your estimate?

➤ Find a way to measure the radius of your sphere.

➤ The volume of a sphere depends on the cube of its radius, r. For your sphere, calculate $r \times r \times r$, or r^3.

Reflect

➤ Divide the volume, V, by r^3.

➤ Compare results with several other pairs of classmates. What do you notice about your values of $V \div r^3$?

Connect the Ideas

An old tennis ball has the same diameter as a frozen juice can. This coincidence can be used to relate the volume of the ball to the volume of the can.

➢ Two matching empty cans are cut so their heights equal the diameter of the ball.

➢ One can is filled with water.

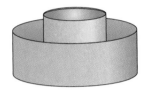

➢ The ball is soaked with water then pushed into the can right to the bottom. Water overflows into a tray.

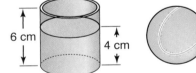

➢ The volume of water in the tray equals the volume of the ball. This water is poured into the other can. The height of the water is measured.

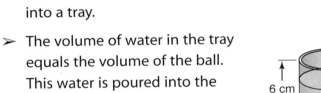

➢ The height of the water is 4 cm. The height of the can is 6 cm.

Since the cans are congruent, the ratio of the volumes of water is equal to the ratio of the heights.

Since 4 cm is $\frac{2}{3}$ of 6 cm, the volume of the ball is $\frac{2}{3}$ the volume of the can.

> Volume of a sphere = $\frac{2}{3}$ × Volume of a cylinder into which the sphere just fits

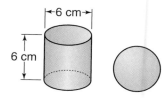

The height of this cylinder equals its diameter.
The volume of the cylinder is about 170 cm³.
The volume of the sphere with the same diameter is about:
$\frac{2}{3} \times 170$ cm³ ≑ 113 cm³

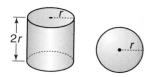

The volume of this cylinder is $\pi r^2 \times 2r = 2\pi r^3$
The volume of the sphere is:
$\frac{2}{3} \times 2\pi r^3 = \frac{4}{3} \pi r^3$

The volume V of a sphere with radius r is: $V = \frac{4}{3} \pi r^3$

Practice

1. The height of this cylinder is twice its radius.
Determine the volumes of the cylinder and the sphere.

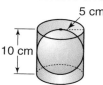

2. The height of this cylinder is twice its radius.
Determine the volume of the sphere.

3. Determine the volume of each sphere.

a) 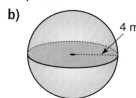 b) 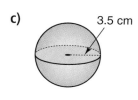 c) 3.5 cm

Many objects are approximately spherical.
Their volumes can be estimated using the formula for the volume of a sphere.

Example

An orange is approximately spherical.
Its diameter is 10 cm.
What is the volume of the orange?

Solution To determine the volume of the orange, use: $V = \frac{4}{3}\pi r^3$
$r = \frac{10 \text{ cm}}{2} = 5$ cm. Substitute: $r = 5$
$V = \frac{4}{3} \times \pi \times 5^3$
$V \doteq 523.599$

The volume of the orange is about 524 cm³.

4. Determine the volume of each sphere.

   a)   b) 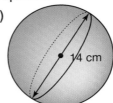   c)

5. An inflated balloon approximates a sphere with radius 11.5 cm. A student's lung capacity is 3.6 L.
   a) How many breaths does the student use to inflate the balloon? What assumptions did you make?
   b) How do you know your answer is reasonable?

6. **Assessment Focus** Lyn has a block of wood that measures 14 cm by 14 cm by 14 cm. She is making a wooden ball in tech class.
   a) What is the volume of wood in the block?
   b) What is the largest possible diameter for the ball?
   c) What is the volume of the wooden ball?
   d) What volume of wood is cut off the block to make the ball? What assumptions did you make?

7. **Take It Further** Meighan is selling ice-cream cones at the fall fair. Each carton of ice cream is 20 cm by 11 cm by 24 cm. The ice-cream scoop makes a sphere of ice cream, with diameter 8 cm.
   a) How many scoops should Meighan get from each carton?
   b) Each carton of ice cream costs $4.29. How much does each scoop cost?
   c) Meighan pays $1.99 for a package of 12 sugar cones. Suggest a price Meighan should charge for each single-scoop and double-scoop cone. Justify your answer.

How is the volume of a sphere related to the volume of a cylinder? Include diagrams in your explanation.

Chapter Review

What Do I Need to Know?

Pythagorean Theorem

$a^2 + b^2 = c^2$, where a and b are the legs of a right triangle and c is the hypotenuse

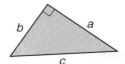

Volume Formulas

Volume of a prism is base area times height.
$V = Bh$

Volume of a cylinder is base area times height.
$V = \pi r^2 h$

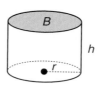

See the Expanded Glossary for the formulas for areas and perimeters of figures.

Volume of a pyramid is one-third the base area times the height.
$V = \frac{1}{3} Bh$

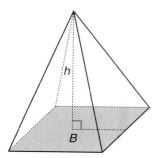

Volume of a cone is one-third the base area times the height.
$V = \frac{1}{3} \pi r^2 h$

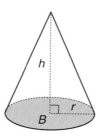

Volume of a sphere is two-thirds the volume of a cylinder into which the sphere just fits.

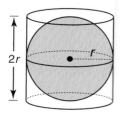

Volume of a sphere is four-thirds π times the cube of the radius.
$V = \frac{4}{3} \pi r^3$

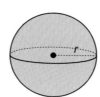

What Should I Be Able to Do?

1.1

1. Determine the perimeter and area of each figure.

a)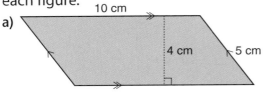

b)

2. A hurricane warning flag is a square with side length 90 cm. The red border is 12 cm wide.

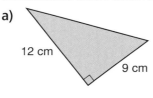

a) What is the area of black material?
b) What is the area of red material?

1.2

3. Determine each unknown length.

a)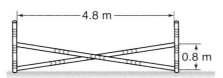

b)

4. Jean set up cross poles for his horse to jump. How long is each cross pole?

1.3

5. Determine the area of this figure. The curve is a semicircle.

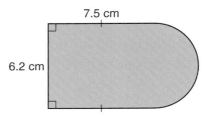

6. The sailing regatta committee has this flag to show a fourth place finish. What is the area of the red material in the flag?

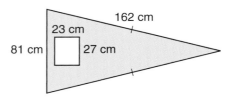

1.4

7. Determine the perimeter of this figure.

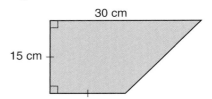

8. A large greeting card has the shape of a square, with a semicircle on each of two sides.

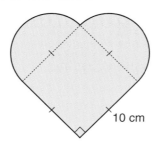

a) There is a ribbon around the perimeter of the card. How long is this ribbon?
b) How do you know your answer is reasonable?

1.5 9. Determine the volume of each object.
a)

b)
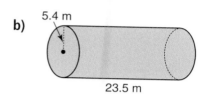

10. A tray of lasagna for 4 people is 19 cm wide, 24 cm long, and 7 cm deep.
    a) Suppose the length and width of the tray are doubled. How many people should this new tray feed? Explain your answer. Include a diagram.
    b) Suppose each dimension of the tray is doubled. How many people should the larger tray feed? Justify your answer.

1.6 11. Determine the volume of the pyramid.
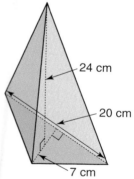

12. The Katimavik Pavilion at Expo '67 is a huge square pyramid. Its base is 20.0 m by 20.0 m. Its height is 14.1 m.
    a) What is the volume of the pyramid?
    b) How do you know your answer is reasonable?

1.7 13. a) What is the volume of the cone?
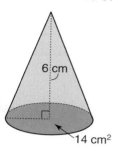
    b) What is the height of this related cylinder?
    c) What would the cylinder's height have to be for it to have the same volume as the cone? Check your answer.

14. Sebastian is filling a conical piñata. How much space is there for candy?
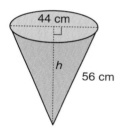

1.8 15. Determine the volume of each sphere.
a) b)
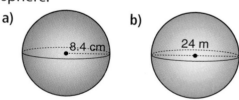

16. In February 2003, Andy Martell of Toronto set a world record for the largest ball of plastic wrap. The ball was approximately spherical. Its diameter was about 43.6 cm.
    a) What was its volume?
    b) How do you know your answer is reasonable?

Chapter Review 39

Practice Test

Multiple Choice: Choose the correct answer for questions 1 and 2.

1. A circle has diameter 16 cm. What is its approximate area?
   A. 631.65 cm²   B. 50.27 cm²   C. 201.06 cm²   D. 804.25 cm²

2. Which statement is true for the volumes of these objects?
   A. All volumes are equal.
   B. The cylinder has the least volume.
   C. Two volumes are equal.
   D. The sphere has the greatest volume.

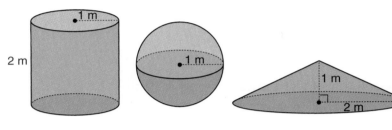

Show all your work for questions 3 to 6.

3. **Knowledge and Understanding** Determine each unknown length.
   a)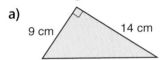
   b)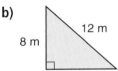

4. **Application** The town works department is ordering materials for a new park. It has the shape of a parallelogram with a right triangle on one side. Here is a plan of the park.
   a) Which measure must you determine to find the area of the park? What is the park's area?
   b) The park is to be fenced except for the gated entrances. There will be four 2.25-m gates in the fence. How much fencing needs to be ordered?
   Justify your answers.

5. **Communication** In △ABC, is ∠ABC a right angle? Justify your answer.

6. **Thinking** Cameron is filling a drink cooler with bags of ice. How many bags of ice are needed to half fill the cooler? What assumptions did you make?

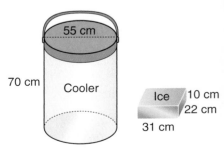

2 Investigating Perimeter and Area of Rectangles

What You'll Learn

To determine the maximum area of a rectangle for a given perimeter and to determine the minimum perimeter of a rectangle for a given area

And Why

In real-world problems, we often want to maximize space or minimize cost.

Key Words
- maximum
- minimum

Project Link

- Build the Biggest Chip Tray

Exploring Guided Examples

All sections of this textbook have one *Guided Example*.
A Guided Example is a detailed solution to a problem.

Reading a Guided Example

Reading a Guided Example will help you learn the math in that section.

When you read a Guided Example, you should make sure you understand how one step follows from the previous steps.

Example A rectangular prism has dimensions 6 m by 4 m by 3 m. Determine the volume of the prism.

Solution Sketch and label a prism.
Use the formula for the volume of a rectangular prism: $V = \ell wh$
Substitute: $\ell = 6, w = 4,$ and $h = 3$
$V = 6 \times 4 \times 3$
$ = 72$
The volume of the prism is 72 m³.

Volume is measured in cubic units.

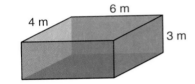

➢ Work with a partner. Read the Solution above.
➢ Make sure you understand each step in the Solution.
➢ Read and explain each step in the Solution to your partner.

Writing a Guided Example

Writing a solution that looks like a Guided Example can help you organize your work and show all your steps.

Use the Guided Example above as a guide.
➢ Write a solution to the following problem.

Example A rectangular prism has dimensions 12 cm by 6 cm by 8 cm. Determine the volume of the prism.

➢ Compare your solution with your partner's.
➢ Do the solutions look like the Guided Example above? Explain any differences.

2.2 Rectangles with Given Perimeter or Area

Investigate Rectangles with a Given Perimeter

Work with a partner.
You will need 0.5-cm grid paper.
Your teacher will give you a loop of string.
How many different rectangles can you make with the loop of string?

➤ Lay the loop of string on grid paper. Place the tips of four pencils inside the loop.

➤ Use the pencils to make the string taut and form a rectangle. Mark the vertices on the grid paper.

Use the lines on the grid paper to make the angles as close to 90° as possible.

➤ Remove the string. Use a ruler to join the vertices and draw the rectangle.
Measure the length and width of the rectangle.
Calculate its area and perimeter.
Record these measures.

Length (cm)	Width (cm)	Area (cm²)	Perimeter (cm)

➤ Use the same loop.
Repeat the previous steps to create 4 different rectangles.

Reflect

➤ How do you know when the figure is a rectangle?

➤ Why do all the rectangles have the same perimeter?

➤ Why do the rectangles have different areas?

Connect the Ideas

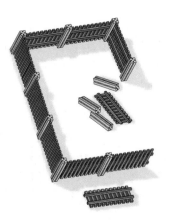

A toy farm set has 20 straight sections of fence.
Each section is 10 cm long.
Here are 2 possible rectangular toy corrals.

1st corral
8 sections long
2 sections wide

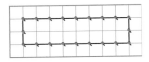

2nd corral
7 sections long
3 sections wide

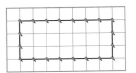

Determine the dimensions

1st corral
$\ell = 8 \times 10$ cm
$ = 80$ cm
$w = 2 \times 10$ cm
$ = 20$ cm
The corral is 80 cm by 20 cm.

2nd corral
$\ell = 7 \times 10$ cm
$ = 70$ cm
$w = 3 \times 10$ cm
$ = 30$ cm
The corral is 70 cm by 30 cm.

We can compare the perimeters of the corrals and their areas.

Determine the perimeters

Use the formula: $P = 2\ell + 2w$

1st corral
Substitute: $\ell = 80$ and $w = 20$
$P = 2(80) + 2(20)$
$ = 160 + 40$
$ = 200$
The perimeter is 200 cm.

2nd corral
Substitute: $\ell = 70$ and $w = 30$
$P = 2(70) + 2(30)$
$ = 140 + 60$
$ = 200$
The perimeter is 200 cm.

Recall that 2(80) means 2×80.

Determine the areas

Use the formula: $A = \ell w$

1st corral
Substitute: $\ell = 80$ and $w = 20$
$A = (80)(20)$
$ = 1600$
The area is 1600 cm^2.

2nd corral
Substitute: $\ell = 70$ and $w = 30$
$A = (70)(30)$
$ = 2100$
The area is 2100 cm^2.

The corrals have the same perimeter but different areas.

Practice

1. a) Draw each rectangle on grid paper.
   i) $\ell = 8, w = 2$ ii) $\ell = 8, w = 3$ iii) $\ell = 6, w = 4$ iv) $\ell = 9, w = 2$
   b) Which rectangles in part a have the same perimeter? How do you know?
   c) Which rectangles in part a have the same area? How do you know?
   d) Can rectangles with the same perimeter have different areas? Explain.
   e) Can rectangles with the same area have different perimeters? Explain.

Example

Jacy wants to frame these pictures.
a) Show that both pictures have the same area.
b) Is the same length of frame needed for each picture? Explain.

Solution   a) Use the formula: $A = \ell w$

Picture A
Substitute: $\ell = 36$ and $w = 25$
$A = (36)(25)$
$\quad = 900$
Both pictures have area 900 cm².

Picture B
Substitute: $\ell = 45$ and $w = 20$
$A = (45)(20)$
$\quad = 900$

Calculate the perimeter of each picture to determine the length of frame.

b) Use the formula: $P = 2\ell + 2w$

Picture A
Substitute: $\ell = 36$ and $w = 25$
$P = 2(36) + 2(25)$
$\quad = 122$
Picture A needs 122 cm of frame.
So, picture B needs more frame.

Picture B
Substitute: $\ell = 45$ and $w = 20$
$P = 2(45) + 2(20)$
$\quad = 130$
Picture B needs 130 cm of frame.

2. a) Draw this rectangle on grid paper.
   Draw 5 different rectangles with the same perimeter as this rectangle.
   b) Calculate the area of each rectangle.
   Write the area inside each rectangle.

   $A = \ell w$

   c) Cut out the rectangles. Sort them from least to greatest area.
   d) Describe how the shapes of the rectangles change as the area increases.

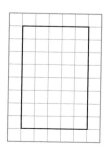

3. a) Draw this rectangle on grid paper.
Draw 3 different rectangles with the same area as this rectangle.

b) Calculate the perimeter of each rectangle.
Write the perimeter inside each rectangle.

$P = 2\ell + 2w$

c) Cut out the rectangles. Sort them from least to greatest perimeter.

d) Describe how the rectangles change as the perimeter increases.

4. Assessment Focus

a) Draw a square on grid paper.
Determine its area and perimeter.

b) Draw a rectangle with the same perimeter but a different area.

c) Which figure has the greater area? Show your work.

5. This diagram shows overlapping rectangles drawn on 1-cm grid paper.
All of them share the same vertex.

a) Calculate the area of each rectangle.
What do you notice?

b) Which rectangle do you think has the least perimeter?
The greatest perimeter? Justify your choice.

c) Calculate the perimeter of each rectangle.
Were you correct in part b? Explain.

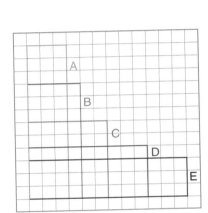

6. This diagram shows overlapping rectangles drawn on 1-cm grid paper. All of them share the same vertex.

a) Calculate the perimeter of each rectangle.
What do you notice?

b) Which rectangle do you think has the least area?
The greatest area? Justify your choice.

c) Calculate the area of each rectangle.
Were you correct in part b? Explain.

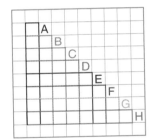

7. Take It Further Can two different rectangles have the same perimeter and the same area? Justify your answer.

In Your Own Words

Explain how rectangles can have the same perimeter but different areas.
Explain how rectangles can have the same area but different perimeters.
Include diagrams in your explanations.

2.3 Maximum Area for a Given Perimeter

A building supply store donates fencing to a day-care centre.
The fencing will be used to create a rectangular play area.
Which dimensions will give the largest possible area?

Investigate Greatest Area of a Rectangle

Work in a group of 4.
You will need a 10 by 10 geoboard and geobands.

➤ Choose one of these perimeters:
10 units, 16 units, 18 units, or 20 units

➤ Make as many different rectangles as you can with that perimeter.
Record the length and width of each rectangle in a table.
Determine the area of each rectangle.

Length (units)	Width (units)	Area (square units)

➤ Which rectangle has the greatest area?

Reflect

➤ How do you know you found the rectangle with the greatest area?

➤ What is the greatest area?
Describe the rectangle with the greatest area.

➤ Share results with your group.
Compare the rectangles with the greatest area for a given perimeter.
How are they the same? How are they different?

Connect the Ideas

Maya has 18 m of edging to create a rectangular flowerbed. What dimensions will give the greatest area?

Since the edging cannot be cut, the dimensions of the rectangles must be whole numbers.

➤ Suppose the edging is in 1-m sections and cannot be cut. Draw rectangles to represent flowerbeds with perimeter 18 m. Calculate the area of each rectangle.

Length (m)	Width (m)	Area (m²)
8	1	8
7	2	14
6	3	18
5	4	20

From the table, the greatest or **maximum** area is 20 m².

The greatest area occurs when the dimensions of the flowerbed are closest in value: 5 m and 4 m

➤ If the edging can be cut, Maya can make flowerbeds with dimensions that are even closer in value.

Length (m)	Width (m)	Area (m²)
4.9	4.1	20.09
4.8	4.2	20.16
4.7	4.3	20.21
4.6	4.4	20.24
4.5	4.5	20.25

From the table, the maximum area is 20.25 m².

This maximum area occurs when the dimensions of the flowerbed are equal.

That is, the rectangle is a square with side length 4.5 cm.

This result is true in general. Among all rectangles with a given perimeter, a square has the maximum area.

It may not be possible to form a square when the sides of the rectangle are restricted to whole numbers.
Then, the maximum area occurs when the length and width are closest in value.

54 CHAPTER 2: Investigating Perimeter and Area of Rectangles

Practice

1. a) On 1-cm grid paper, draw all possible rectangles with each perimeter.
  Each dimension is a whole number of centimetres.
  i) 8 cm   **ii)** 12 cm   **iii)** 22 cm
 b) Calculate the area of each rectangle in part a.
 c) For each perimeter in part a, what are the dimensions of the rectangle with the maximum area?

2. Suppose you have 14 sections of fence to enclose a rectangular garden.

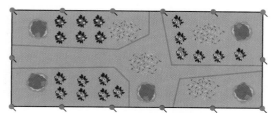

Each section is 1 m long and cannot be cut.
 a) Sketch the possible gardens that can be enclosed by the fence. Label each garden with its dimensions.
 b) Calculate the area of each garden in part a.
 c) Which garden has the maximum area? Is this garden a square? Explain.

Example

What is the maximum area of a rectangle with perimeter 30 m?

Solution

The maximum area occurs when the rectangle is a square.
Determine the side length, s, of a square with perimeter 30 m.
Use the formula: $P = 4s$
Substitute: $P = 30$
$30 = 4s$
Think: What do we multiply 4 by to get 30?
Divide 30 by 4 to find out.
$s = \frac{30}{4}$
$s = 7.5$
Calculate the area of the square. Use the formula: $A = s^2$
Substitute: $s = 7.5$
$A = (7.5)^2$
$ = 56.25$
The maximum area is 56.25 m².

3. What is the maximum area for a rectangle with each perimeter? Explain your thinking.
   a) 36 cm
   b) 60 cm
   c) 75 cm

4. **Assessment Focus** Determine the dimensions of a rectangle with perimeter 42 m whose area is as great as possible. What is the maximum area? Justify your answer.

Need Help?
Read the Example on page 55.

5. Joachim has been comparing rectangles that have the same perimeter. He says that rectangles whose lengths and widths are close in value have larger areas than rectangles whose lengths and widths are very different. Do you agree with Joachim? Justify your answer. Include diagrams and calculations.

6. In a banquet room, there are small square tables that seat 1 person on each side. These tables are pushed together to create a rectangular table that seats 20 people.
   a) Consider all possible arrangements of square tables to seat 20 people. Sketch each arrangement.
   b) Which arrangement requires the most tables? The fewest tables?
   c) Explain why the arrangement with the fewest tables might not be preferred in this situation.

Three tables seat 8 people.

Which tools could you use to help you solve this problem?

7. **Take It Further** Lisa and her sister Bonnie are each given 8 m of plastic edging to create a flowerbed. Lisa creates a square flowerbed. Bonnie creates a circular flowerbed.
   a) Calculate the area of each girl's flowerbed.
   b) Which figure encloses the greater area? Do you think the areas of all squares and circles with the same perimeter are related this way? Draw other squares and circles to find out.

Use *The Geometer's Sketchpad* if available.

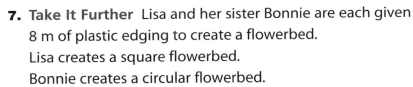

In Your Own Words

Suppose your friend was absent during this lesson. Describe what you learned. Include diagrams and examples in your explanation.

Using *The Geometer's Sketchpad* to Investigate Maximum Area and Minimum Perimeter

Part 1: Maximum Area

1. Open the file *MaxArea.gsp*.

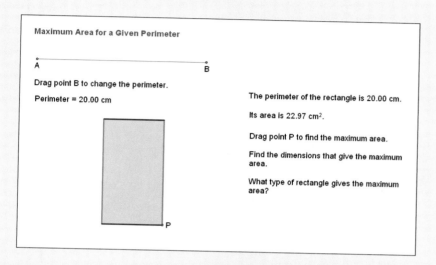

2. The rectangle in the sketch has perimeter 20 cm. Drag point P to see how the area of the rectangle varies while the perimeter stays the same.
   a) What is the maximum area of the rectangle?
   b) Measure the length and width of the rectangle with maximum area. What do you notice?
   c) Record your results in a table.

Your teacher will tell you how to measure with *The Geometer's Sketchpad*.

Perimeter (cm)	Maximum area (cm²)	Length (cm)	Width (cm)
20			

3. a) Drag point B to create a rectangle with a different perimeter.
   b) Determine the maximum area of the rectangle.
   c) Record your results in the table.

4. Repeat the steps in question 3 for 3 more rectangles.

5. Describe the patterns in the table.

Technology: Using *The Geometer's Sketchpad* to Investigate Maximum Area and Minimum Perimeter **57**

Part 2: Minimum Perimeter

1. Open the file *MinPerim.gsp*.

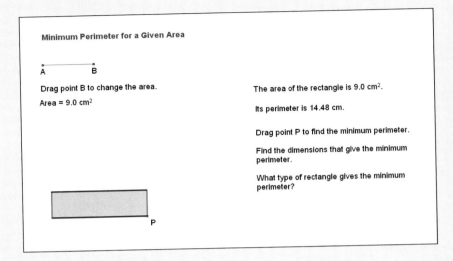

2. The rectangle in the sketch has area 9 cm². Drag point P to see how the perimeter of the rectangle varies while the area stays the same.
   a) What is the minimum perimeter of the rectangle?
   b) Measure the length and width of the rectangle with minimum perimeter.
   What do you notice?
   c) Record your results in a table.

Area (cm²)	Minimum perimeter (cm)	Length (cm)	Width (cm)
9			

3. a) Drag point B to create a rectangle with a different area.
   b) Determine the minimum perimeter of the rectangle.
   c) Record your results in the table.

4. Repeat the steps in question 3 for 2 more rectangles.

5. Describe the patterns in the table.

Find the Least Perimeter

Play with a classmate.

➤ Player B deals 6 cards to each player.
He places the other cards face down.

➤ Player A rolls the number cube.
The number rolled represents the area of a rectangle.

➤ Player B finds 2 cards that are the length and width
of a rectangle with that area.
Player B states the perimeter of the rectangle.
If he is correct, he gets 1 point.
If it is the least perimeter for that rectangle, he gets 2 points.
Player B puts the cards in a discard pile and takes 2 more cards
from the remaining deck.
Then Player B rolls the cube for Player A to find the cards.

➤ If Player B does not have the 2 cards, he can take a card from
the remaining deck.
If Player B still does not have the 2 cards he needs,
Player A tries to find the 2 cards.
If no one has the two cards, Player A rolls the cube again.

➤ Play continues until no cards remain in the deck.
The person with more points wins.

Materials:

- deck of playing cards

A = 1   Jack = 12

Queen = 14   King = 16

- number cube labelled 16, 20, 24, 28, 32, 36

The numbers on the cube are the areas of rectangles.

Mid-Chapter Review

2.1 1. Explain how the area and perimeter of the rectangle changes in each scenario.

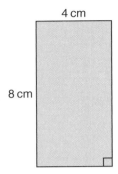

a) The length stays the same but the width increases.
b) The length stays the same but the width decreases.
c) Both the length and width increase.
d) Both the length and width decrease.
e) The length decreases and the width increases.
The amount of decrease equals the amount of increase.

2. Alat made these rectangles on a geoboard.

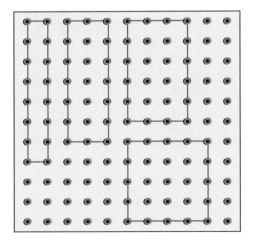

a) What is the same about all these rectangles?
b) What is different?

2.2 3. a) List 4 possible lengths and widths for a rectangle with each area.
  i) 30 m²   ii) 64 m²
b) Determine the perimeter of each rectangle in part a.

4. A rectangle is 15 cm long and 9 cm wide. Its length and width change as described below.
a) The length increases by 3 cm and the width decreases by 3 cm.
b) The length is multiplied by 3 and the width is divided by 3.
In each case, does the area change? Does the perimeter change? Explain. Include diagrams and calculations in your explanation.

5. A farmer has 2 horse paddocks. One paddock is 10 m by 90 m. The other paddock is 30 m by 30 m.
a) Show that both paddocks have the same area.
b) Is the same amount of fencing required to enclose each paddock? Explain.

2.3 6. Draw these rectangles on grid paper.
Rectangle A: 20 cm by 10 cm
Rectangle B: Same perimeter as rectangle A but smaller area
Rectangle C: Same perimeter as rectangle A and the greatest possible area
Write the dimensions, perimeter, and area of each rectangle.

7. What is the maximum area of a rectangular lot that can be enclosed by 180 m of fencing?

2.4 Minimum Perimeter for a Given Area

Miriam works in a garden centre.
She is asked what the least amount of edging would be for a certain rectangular area of grass.
How might she find out?

Investigate Least Edging for a Patio

Michael has 36 square stones to arrange as a rectangular patio.
He will then buy edging to go around the patio.

➤ Sketch the different patios Michael can create.
  Label each patio with its dimensions.

Which tools could you use to help you solve this problem?

➤ Which patio requires the least amount of edging?

➤ Suppose each stone has side length 30 cm.
  What is the least amount of edging needed for the patio?
  Justify your answer.

Reflect

➤ Is a 9 by 4 arrangement the same as a 4 by 9 arrangement? Explain.

➤ Which measure is the same for all the patios?
  Which measures are different?

➤ What assumptions or estimates did you make to calculate the amount of edging?

➤ Describe the rectangle that uses the least amount of edging.

Connect the Ideas

➢ Eric plans to build a storage shed with floor area 60 m².
He wants to use the least amount of materials.
So, Eric needs the least perimeter for the floor.
He will use 60 square patio stones, each with area 1 m².

Here are some rectangles that represent floors with area 60 m².

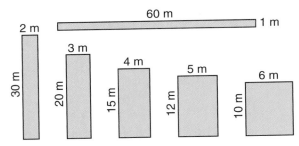

The rectangles are not drawn to scale.

Here are some possible dimensions and perimeters of the floor.

Length (m)	Width (m)	Perimeter (m)
60	1	122
30	2	64
20	3	46
15	4	38
12	5	34
10	**6**	**32**

From the table, the least or **minimum** perimeter is 32 m.

The minimum perimeter occurs when the dimensions of the floor are closest in value: 10 m and 6 m

We do not need to continue the table because we would repeat the rectangles we have.

➢ Jenna suggests that Eric could have a lesser perimeter if he used concrete instead of patio stones.
Jenna drew a square with area 60 m².

The area of the square is $s \times s$. The area is also 60 m².
To find two equal numbers whose product is 60,
Jenna used her calculator to determine the square root of 60.
$\sqrt{60} \doteq 7.75$
A square with side length 7.75 m has perimeter:
4×7.75 m = 31 m

This is true in general. Among all rectangles with a given area, a square has the minimum perimeter.

It may not be possible to form a square when the sides of the rectangle are restricted to whole numbers.
Then, the minimum perimeter occurs when the length and width are closest in value.

Practice

1. a) Draw Rectangle A on grid paper.
   b) Draw a different rectangle that has the same area as Rectangle A.
   c) Do the two rectangles have the same perimeter? Explain.

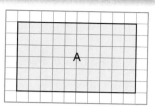

2. a) Draw Rectangle B on grid paper.
   b) Draw a different rectangle that has the same area as Rectangle B, but a lesser perimeter.
   c) Draw a rectangle that has the same area as Rectangle B with the minimum perimeter.

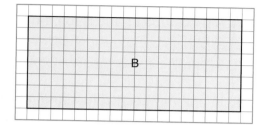

3. a) For each number of congruent square patio stones, which arrangement would give the minimum perimeter?
   i) 25 stones   ii) 50 stones   iii) 75 stones   iv) 100 stones

   Remember that congruent squares are the same size.

   b) Suppose each stone in part a has side length 30 cm. What is the minimum perimeter?

4. Look at the areas and perimeters in question 3.
   a) What patterns do you see in the areas?
   b) Do the perimeters have the same pattern? Explain.

Example   What is the minimum perimeter for a rectangle with area 40 m²?

Solution   The minimum perimeter occurs when the rectangle is a square.
Determine the side length, s, of a square with area 40 m².
Use the formula: $A = s^2$
Substitute: $A = 40$
$40 = s^2$
$s = \sqrt{40}$
$s \doteq 6.325$

Calculate the perimeter of the square.
Use the formula: $A = 4s$
Substitute: $s = 6.325$
$P = 4(6.325)$
$ = 25.3$

The minimum perimeter for the rectangle is approximately 25.3 m.

In questions 5 to 7, round your answers to 1 decimal place.

5. What is the minimum perimeter for a rectangle with each area?
   a) 30 m²   b) 60 m²   c) 90 m²   d) 120 m²

6. **Assessment Focus** Determine the dimensions of a rectangle with area 1000 m² whose perimeter is the least possible. What is the minimum perimeter? Justify your answer.

Having Trouble?
Read Connect the Ideas.

7. A rectangular patio is to be built from 56 congruent square tiles.
   a) Which patio design will give the minimum perimeter?
   b) Is this patio a square? Explain.

8. Keung has been comparing rectangles that have the same area. He says that rectangles whose lengths and widths are close in value have greater perimeters than rectangles whose lengths and widths are very different. Do you agree? Explain.

9. In a banquet room, there are small square tables that seat 1 person on each side.
   The tables are pushed together to create larger rectangular tables.
   a) Consider all possible arrangements of 12 square tables.
      Sketch each arrangement on grid paper.
   b) Which arrangement seats the most people? The fewest people? Explain.
   c) Explain why the minimum perimeter might not be preferred in this situation.

10. **Take It Further**
    a) Draw a square.
    b) Draw a rectangle that has the same area as the square but a different perimeter.
    c) Is it possible to draw a rectangle that has the same area as the square but a lesser perimeter than the square? Explain.

Which tools could you use to solve this problem?

In Your Own Words

For a given area, describe the rectangle that has the least perimeter. Will the rectangle always be a square? Explain.

64   CHAPTER 2: Investigating Perimeter and Area of Rectangles

2.5 Problems Involving Maximum or Minimum Measures

Recall, from page 61, that Michael was building a rectangular patio with 36 square stones. To save money on edging, Michael decides to place the patio against the side of the house.
Now, edging will be needed on only three sides of the patio.

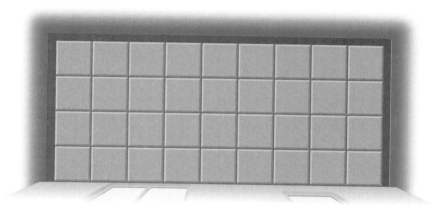

Investigate — Solving a Problem About Minimum Length

For Michael's patio:

Which tools could you use to solve this problem?

➢ What should the dimensions be so the patio uses the minimum length of edging?

➢ Each stone has side length 30 cm. What is the minimum length of edging? Justify your answer.

Reflect

➢ Could you have used the patios you drew in Section 2.4 *Investigate*? Explain.

➢ Will a 4 by 9 patio and a 9 by 4 patio require the same length of edging? Explain.

➢ Is the patio that requires the minimum length of edging a square? Explain.

Connect the Ideas

A rectangular swimming area is to be enclosed by 100 m of rope.
One side of the swimming area is along the shore,
so the rope will only be used on three sides.
What is the maximum swimming area that can be created?

Sketch rectangles to represent the swimming areas.
Calculate the area of each rectangle.

For example, consider a swimming area with width 5 m.

Two widths and 1 length are formed by the rope.
The rope is 100 m.
Two widths are: 2(5 m) = 10 m
So, the length is: 100 m − 10 m = 90 m
The area is: (5 m)(90 m) = 450 m²

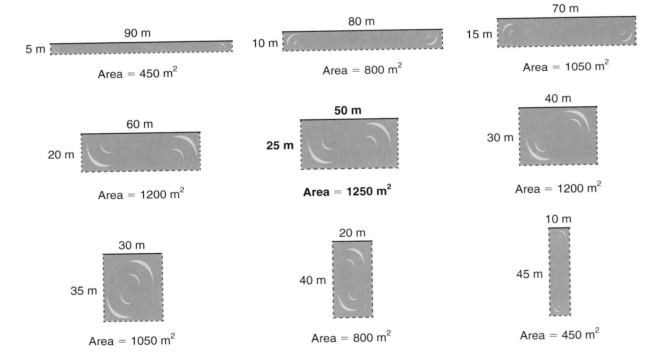

The areas increase, then decrease.
The maximum swimming area is 1250 m².
Only one of the rectangles has this area.

The maximum swimming area is not a square.
It is a rectangle that measures 50 m by 25 m.

Practice

1. A store owner uses 16 m of rope for a rectangular display. There is a wall on one side, so the rope is only used for the other 3 sides.
   a) Copy these rectangles. Determine the missing measures.

   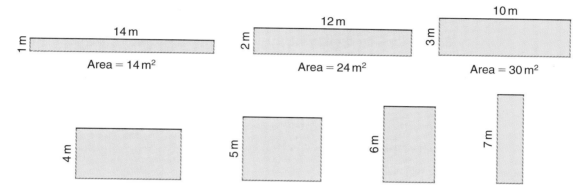

   b) What are the dimensions of the maximum area that can be enclosed? How are these dimensions related?

2. Cody has 20 m of fencing to build a rectangular pen for his dog. One side of the pen will be against his house.
   a) Sketch and label as many possible rectangles as you can.
   b) Copy and complete this table.

Width (m)	Length (m)	Area (m²)
1	18	18

   Continue until the width is 9 m.

   c) What dimensions give the maximum area for the pen? How are these dimensions related?

3. **Assessment Focus** A rectangular field is enclosed by 80 m of fencing.
   a) What are the dimensions of the field with maximum area in each situation?

   i) The fence encloses the entire field.

   ii) One side of the field is not fenced.

   b) How are the dimensions of the rectangles in part a related?

2.5 Problems Involving Maximum or Minimum Measures

Example A patio is to be built on the side of a house using 24 congruent square stones. It will then be edged on 3 sides. Which arrangement requires the minimum edging?

Solution The area of the patio is 24 square units. List pairs of numbers with product 24. Calculate the edging required for each patio.
For a 3 by 8 arrangement, this is:
Amount of edging = 2(3) + 8
                 = 14
A 4 by 6 arrangement uses the same amount of edging. So, there are 2 different arrangements that require the minimum edging.

Width (units)	Length (units)	Edging (units)
1	24	26
2	12	16
3	8	14
4	6	14
6	4	16
8	3	19
12	2	26
24	1	49

4. A patio of 72 congruent square stones is to be built on the side of a house. Sketch and label possible patio designs. Which design has the minimum distance around the sides not attached to the house?

5. A patio of 72 congruent square stones is to be built in the corner of a backyard against a fence. Sketch and label possible patio designs. Which design requires the minimum edging for the sides not against the fence?

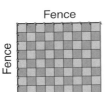

6. Take It Further Twenty-six sections of fence, each 1 m long, are used to create two storage enclosures of equal area.
  a) Find another way to arrange 26 sections to make two equal storage areas.
  The storage areas should share one side.
  b) What should the length and width of each storage area be to make the enclosure with the maximum area? What are the dimensions of this enclosure?

Which tools could you use to solve this problem?

In Your Own Words

For a given area, you have found the rectangle with the minimum perimeter and the rectangle with the minimum length of 3 sides. How are the rectangles different? Include sketches in your explanation.

Chapter Review

What Do I Need to Know?

Area and Perimeter Formulas

Rectangle

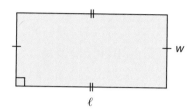

Area: $A = \ell w$
Perimeter: $P = 2(\ell + w)$
or, $P = 2\ell + 2w$

Square

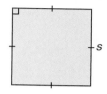

Area: $A = s^2$
Perimeter: $P = 4s$

Maximum Area of a Rectangle

Rectangles with the same perimeter can have different areas.
For example, all these rectangles have perimeter 16 cm.

If the dimensions of the rectangle are restricted to whole numbers, it may not be possible to form a square. The maximum area occurs when the length and width are closest in value.

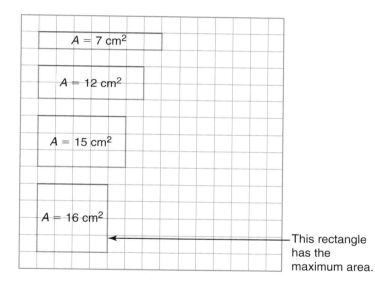

This rectangle has the maximum area.

Minimum Perimeter of a Rectangle

Rectangles with the same area can have different perimeters.
For example, all these rectangles have area 16 m².

If the rectangle is made from congruent square tiles, it may not be possible to form a square.
The minimum perimeter occurs when the length and width are closest in value.

This rectangle has the minimum perimeter.

What Should I Be Able to Do?

2.1

1. A 3-m ladder leans against a wall. The ladder slips down the wall.

   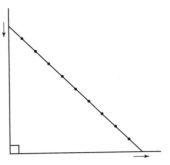

   a) Which measures stay the same?
   b) Which measures vary? Tell whether each measure is increasing or decreasing.

2.2

2. A rectangle has perimeter 100 cm.
   a) Copy and complete this table. For each width, determine the length and area of the rectangle.

Width (cm)	Length (cm)	Area (cm²)
5		
10		
15		
20		
25		

   b) Describe the patterns in the table.

3. A rectangle has area 72 cm².
   a) Copy and complete this table. For each width, determine the length and perimeter of the rectangle.

Width (cm)	Length (cm)	Perimeter (cm)
2		
4		
6		
8		

   b) Describe the patterns in the table.

4. a) Determine the perimeter and area of square B.

   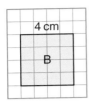

   b) Draw a rectangle that has the same perimeter as square B, but a different area.
   c) Can you draw a rectangle that has the same perimeter as square B, but a greater area? Justify your answer.

2.3

5. a) Do rectangle OABC and rectangle ODEF have the same perimeter? The same area? Explain.

   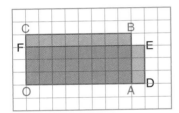

   b) Copy the rectangles in part a on grid paper. On the same diagram, draw as many rectangles as you can that have the same perimeter as rectangle OABC.
   c) Describe the patterns in the rectangles.
   d) In part b, describe the rectangle with:
      i) the maximum area
      ii) the minimum area
   e) Suppose you drew the rectangles in part b on plain paper. Would your answers to part d change? Explain.

6. a) For each perimeter below, determine the dimensions of the rectangle with the maximum area.
   i) 8 cm   ii) 34 cm   iii) 54 cm
   b) Calculate the area of each rectangle in part a. How do you know each area is a maximum?

7. Steve has 58 m of fencing for a rectangular garden.
   a) What are the dimensions of the largest garden Steve can enclose?
   b) What is the area of this garden?

8. Refer to question 7. Suppose the fencing comes in 1-m lengths that cannot be cut.
   a) What are the dimensions of the largest garden Steve can enclose?
   b) What is the area of this garden?

2.4

9. a) For each area below, determine the dimensions of a rectangle with the minimum perimeter.
   i) 16 m²   ii) 24 m²   iii) 50 m²
   b) Calculate the perimeter of each rectangle in part a. How do you know each perimeter is a minimum?

10. Determine the dimensions of a rectangle with area 196 cm² and the minimum perimeter. What is the minimum perimeter?

11. Rachel plans to build a rectangular patio with 56 square stones.
   a) Sketch the different patios Rachel can create. Label each patio with its dimensions.
   b) Suppose Rachel decides to put edging around the patio. Which patio requires the minimum amount of edging? Justify your answer.

2.5 **12.** A lifeguard has 400 m of rope to enclose a rectangular swimming area. One side of the rectangle is the beach, and does not need rope.

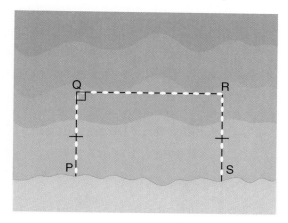

   a) Suppose side QR measures 300 m. What is the length of RS?
   b) Suppose side RS measures 85 m. What is the length of QR?
   c) What are the dimensions of the rectangle with the maximum area? Justify your answer.

Chapter Review **71**

Practice Test

Multiple Choice: Choose the correct answer for questions 1 and 2.

1. Pina makes a rectangular garden surrounded by 32 m of edging. Which of these figures has the maximum area?
   A. A 2-m by 4-m rectangle
   B. A 4-m by 12-m rectangle
   C. A 6-m by 10-m rectangle
   D. A square with side length 8 m

2. A rectangle has an area of 36 m². Which dimensions will ensure that the rectangle has the minimum perimeter?
   A. 6 m by 6 m   B. 9 m by 4 m   C. 12 m by 3 m   D. 18 m by 2 m

Show all your work for questions 3 to 7.

3. **Application** Lindsay was planning to build a rectangular patio with 24 square stones and to edge it on all four sides.
   a) Sketch the different patios Lindsay can create. Label each patio with its dimensions.
   b) Which patio requires the minimum amount of edging? Justify your answer.

4. **Knowledge and Understanding** Determine the maximum area of a rectangle with each given perimeter.
   a) 28 cm
   b) 86 cm

5. **Communication** When you see a rectangle, how do you know if:
   a) it has the minimum perimeter for its area?
   b) it has the maximum area for its perimeter?
   Include diagrams in your explanation.

6. **Thinking** Eighteen sections of fence, each 1 m long, are used to make a rectangular storage area.
   a) How many sections should be used for the length and the width to make the storage area a maximum? What is this area?
   b) Suppose workers found 2 more sections. Where should these sections be put for the maximum storage area? Justify your answer.

7. Suppose the storage area in question 6 is bordered by a wall on one side and 18 sections around the remaining 3 sides. What are the dimensions of the largest area that can be formed?

3 Relationships in Geometry

What You'll Learn

The properties and relationships among angles in polygons and angles involving parallel lines

And Why

These relationships are often used in carpentry and construction; for example, to ensure that floor tiles align correctly or that the sides of a door frame are parallel.

Key Words
- interior angle
- exterior angle
- transversal
- corresponding angles
- alternate angles

Project Link

- Designing with Shape, Colour, and Motion

Using a Frayer Model

Every math topic has words and concepts for you to learn.

A Frayer model is a tool that helps you visualize new words and concepts.
It makes connections to what you already know.

Here is an example of a Frayer model.

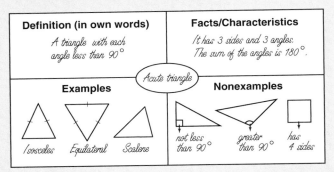

See if you can figure out which word goes in the centre of this Frayer model.

Definition (in own words)	Facts/Characteristics
A figure with 4 sides and 4 angles	The sides are straight. The figure is closed.
Examples	**Nonexamples**
Square Rectangle Parallelogram Trapezoid Kite	Triangle Pentagon Circle Cube

(centre: ?)

Make your own Frayer model for each of these concepts:
- Isosceles triangle
- Pythagorean Theorem

As you work through this chapter, look for ways to use a Frayer model.
Scan each section for new words and new concepts.
Try to make at least 3 Frayer models in this chapter.
Here are some suggestions:
- Exterior angle
- Parallel lines
- Quadrilateral

CHAPTER 3: Relationships in Geometry

3.1 Angles in Triangles

Investigate The Sum of the Interior Angles in a Triangle

Work in a group of 3.
You will need scissors.

Draw *large* triangles.

➤ Use a ruler. Each of you draws one of these triangles: acute triangle, obtuse triangle, and right triangle. Cut out the triangles.

➤ Place each triangle so its longest side is at the bottom and its shortest side is to your left.
Label the angles as shown.
Turn over the triangle so the labels are face down.

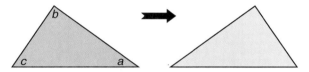

➤ Make a fold as shown.
The crease is at right angles to the bottom side.

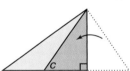

➤ Unfold the triangle and lay it flat.
Mark the point M where the crease meets the bottom side.

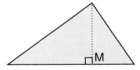

➤ Fold each vertex of the triangle to meet at point M.

➤ How do the three angles of the triangle appear to be related?

Reflect

➤ Compare your results with those of your group.

➤ Did everyone get the same relationship? Explain.

Connect the Ideas

Zahara constructs a triangle using *The Geometer's Sketchpad*. She measures the angles in the triangle and calculates their sum.

The 3 angles in a triangle are its **interior angles**.

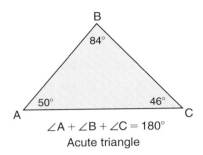

∠A + ∠B + ∠C = 180°
Acute triangle

Zahara drags the vertices to create different types of triangles. The angle measures change, but their sum does not.

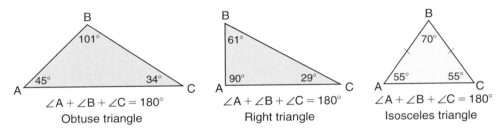

The sum of the three interior angles of a triangle is 180°.
That is, $a + b + c = 180°$

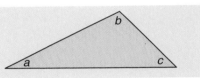

We can use this relationship to determine the third angle in a triangle when we know the other two angles.

In △ABC, ∠A + ∠B + ∠C = 180°

∠B + ∠C = 36° + 46°
   = 82°

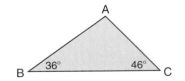

To determine ∠A, think:
What do we add to 82° to get 180°?
Subtract to find out.
180° − 82° = 98°

So, ∠A = 98°

76 CHAPTER 3: Relationships in Geometry

Practice

1. Use a ruler to draw a large triangle.
   Cut out the triangle and tear off its corners.
   Put the corners together as shown.
   Which relationship does this illustrate? Explain.

   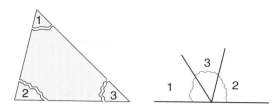

2. Determine each unknown angle.

   a) 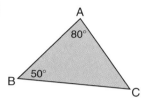   b) 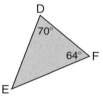   c)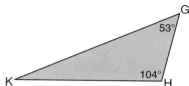

3. For each set of angles, can a triangle be drawn?
   Justify your answers.
   a) 41°, 72°, 67°   b) 40°, 60°, 100°   c) 100°, 45°, 30°

4. Can a triangle have two 90° angles? Explain.

5. **Assessment Focus**
   a) Draw then cut out a right triangle. Fold the triangle as shown. What does this tell you about the acute angles of a right triangle?
   b) Use what you know about the sum of the angles in a triangle to explain your result in part a.

   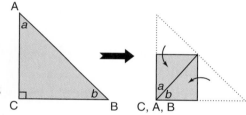

   c) Determine each unknown angle. Explain how you did this.

   i) 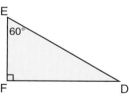   ii) 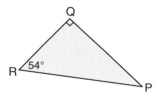   iii)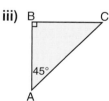

6. Draw then cut out an isosceles triangle.
   Fold the triangle as shown.
   What does this tell you about the angles opposite the equal sides of an isosceles triangle?

   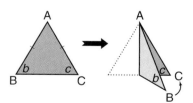

   Example
   a) Write a relationship for the measures of the angles in isosceles △DEF.
   b) Use the relationship to determine the measures of ∠D and ∠F.

   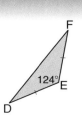

3.1 Angles in Triangles

Solution a) The sum of the angles in a triangle is 180°.
So, ∠D + ∠F + 124° = 180°

b) ∠D + ∠F = 180° − 124°
   = 56°

In an isosceles triangle, the angles opposite the equal sides are equal.
DE = FE, so ∠D = ∠F
There are 2 equal angles with a sum of 56°.
So, each angle is: $\frac{56°}{2}$ = 28°

∠D = 28° and ∠F = 28°

> Think: What do we add to 124° to get 180°? Subtract to find out.

> Recall that $\frac{56°}{2}$ means 56° ÷ 2.

7. In each isosceles △ABC below:
   i) Write a relationship for the measures of the angles.
   ii) Use the relationship to determine the measures of ∠A and ∠B.

   a)
   b)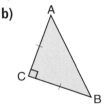
   c)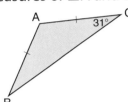
   d)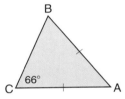

8. Most cars have a scissors jack. It has all 4 sides equal.
   a) The angle shown is 40°. Determine the angle at the top of the jack.
   b) Suppose the angle shown increases to 42°. What happens to the angle at the top of the jack? Explain how you know.

9. An equilateral triangle has 3 equal sides and 3 equal angles. Use this information to determine the measure of each angle. Show your work.

10. **Take It Further** Determine the angle measure indicated by each letter.
    a)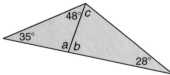
    b)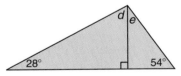
    c)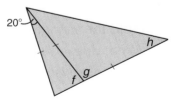

In Your Own Words

Suppose you know the measure of one angle in a triangle. What else would you need to know about the triangle to determine the measures of all its angles? Explain.

Using *The Geometer's Sketchpad* to Investigate Angles Related to Triangles

Part 1: The Relationship among the Interior Angles

1. Open the file *TriangleAngles.gsp*. Click the tab for page 1.

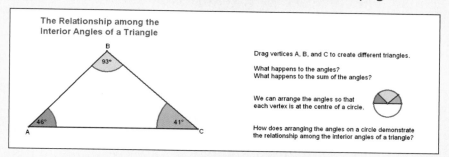

2. Drag vertices A, B, and C to create 5 different triangles.
   a) For each triangle, copy this table and record the measures of ∠A, ∠B, and ∠C.

∠A	∠B	∠C	∠A + ∠B + ∠C

   b) Calculate the sum of each set of angles.
   c) How do the interior angles of a triangle appear to be related?

3. We can arrange ∠A, ∠B, and ∠C so the vertex of each angle is at the centre of a circle.
   a) Drag vertices A, B, and C to create different triangles. Observe how the angles change on the circle.
   b) How does arranging the angles on a circle support your answer to question 2c?

Part 2: The Relationship among the Interior and Exterior Angles

1. Use the file *TriangleAngles.gsp*. Click the tab for page 2.

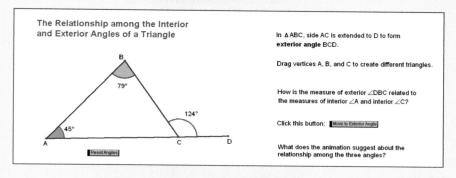

2. In △ABC, side AC is extended to D to form exterior ∠BCD.
   a) Drag vertices A, B, and C to create 5 different triangles.
      For each triangle, copy this table and record the measures of ∠A, ∠B, and ∠BCD.

∠A	∠B	∠BCD

   b) How do the three angles appear to be related? Explain why.
   c) Click the button: *Move to Exterior Angle*
      How does this animation support your answer to part b?

Part 3: The Relationship among the Exterior Angles

1. Use the file *TriangleAngles.gsp*. Click the tab for page 3.

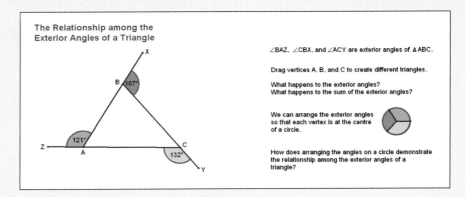

2. Drag vertices A, B, and C to create 5 different triangles.
   a) For each triangle, copy this table and record the measures of the exterior angles.

∠BAZ	∠CBX	∠ACY	∠BAZ + ∠CBX + ∠ACY

   b) Calculate the sum of each set of angles.
   c) How do the exterior angles of a triangle appear to be related? Explain why.

3. We can arrange ∠BAZ, ∠CBX, and ∠ACY so each vertex is at the centre of a circle.
   a) Drag vertices A, B, and C to create different triangles. Observe how the angles change on the circle.
   b) How does arranging the angles on a circle support your answer to question 2c?

3.2 Exterior Angles of a Triangle

Enrico is a carpenter. He is building a ramp. Enrico knows that the greater angle between the ramp and the floor is 160°. How can Enrico calculate the angle the vertical support makes with the ramp?

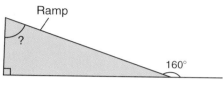

An exterior angle is formed outside a triangle when one side is extended.
Angle GBC is an exterior angle of △ABC.

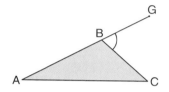

Investigate — Relating Exterior and Interior Angles of a Triangle

Work in a group of 3.

➢ Each of you draws one of these triangles:
  – acute triangle
  – right triangle
  – obtuse triangle

➢ Extend one side to form an exterior angle.
  Measure and record the exterior angle and the interior angles.
  Look for relationships between the exterior angle and one or more of the interior angles.

Reflect

Compare your results with those of your group.

➢ How is the exterior angle related to the interior angle beside it?

➢ How is the exterior angle related to the other two interior angles?

➢ Do your results depend on the type of triangle? Explain.

➢ Explain your results using facts you already know about angles and their sums.

Connect the Ideas

Exterior and interior angles

Jamie cuts out a paper triangle. He tears off the corners at A and B and finds that they fit exterior ∠ACD. That is, ∠ACD = ∠A + ∠B

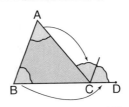

> Each exterior angle of a triangle is equal to the sum of the two opposite interior angles.
> That is, $e = a + b$
>
>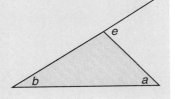

We can use this relationship to calculate the measure of ∠A.
∠ACD = ∠A + ∠B
So, 135° = ∠A + 93°
Think: What do we add to 93° to get 135°?
Subtract to find out.
135° − 93° = 42°
∠A = 42°

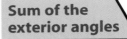

Sum of the exterior angles

Extend the sides of △ABC to form the other exterior angles.

∠BAE = 180° − 42°
      = 138°
∠FBC = 180° − 93°
      = 87°

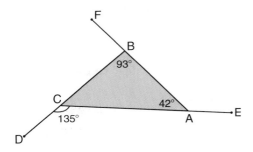

The sum of the exterior angles is:
135° + 138° + 87° = 360°

This relationship is true for any triangle.

> The sum of the 3 exterior angles of a triangle is 360°.
> That is, $p + q + r = 360°$
>
>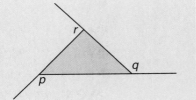

CHAPTER 3: Relationships in Geometry

Practice

1. a) For each exterior angle, name the opposite interior angles.
   b) Write a relationship between the exterior angle and the opposite interior angles.

   i)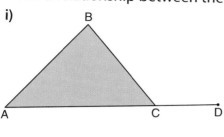
   ii)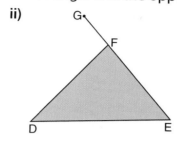
   iii)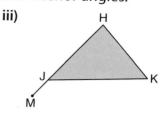

2. Determine each unknown exterior angle. Which relationship are you using each time?

   a)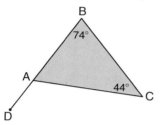
   b)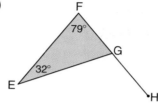
   c)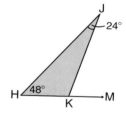

   d)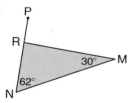
   e)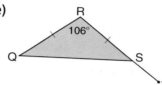
   f)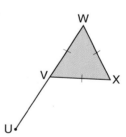

3. Determine the unknown interior angles. Justify your answers.

   a)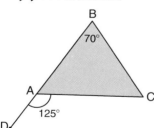
   b)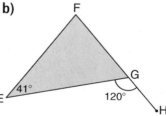
   c)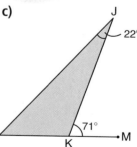

4. A carpenter uses an L-square to measure the timber for a rafter. The ridge angle is 130°. What is the plate angle? How do you know?

   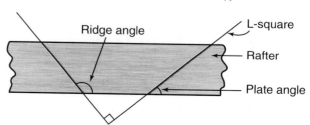

3.2 Exterior Angles of a Triangle   **83**

5. **Assessment Focus** Grace's tech design class builds a model of a roof truss. The broken line is a line of symmetry. Determine the measure of each angle in the truss. Explain how you know.

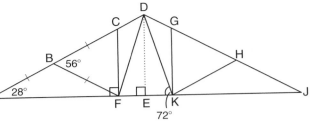

6. Draw a triangle with an exterior angle of 160° and another exterior angle of 72°.
   a) Measure the third exterior angle.
   b) How are the three exterior angles related? Explain.
   c) Compare your triangle and your results with your classmates. What do you notice?

When we know two exterior angles, we can determine the third exterior angle.

Example Determine the angle measure indicated by x.

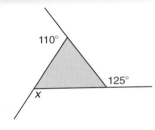

Solution The sum of the 3 exterior angles of a triangle is 360°.
So, $x + 110° + 125° = 360°$
$x + 235° = 360°$
$x = 360° - 235°$
$= 125°$

7. Determine the angle measure indicated by each letter. Justify your answers.

a) 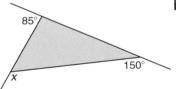   b) 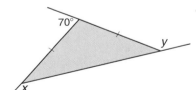   c)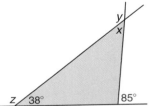

8. **Take It Further** Sketch a triangle with two exterior angles of 150° and 130°. Describe all the different ways you can determine the measures of the third exterior angle and the three interior angles.

In Your Own Words

What do you know about the exterior angles of a triangle? Use diagrams to explain.

Using *The Geometer's Sketchpad* to Investigate Angles Involving Parallel Lines

1. Open the file *Parallel.gsp*. Click the tab for page 1.

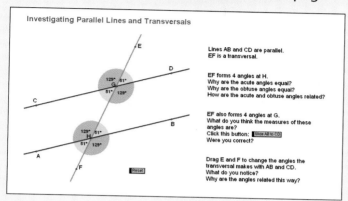

2. Transversal EF intersects AB to form 4 angles at H.
   a) Why are the acute angles equal?
   b) Why are the obtuse angles equal?
   c) How are the acute and obtuse angles related? Explain.

3. CD is parallel to AB. EF intersects CD to form 4 angles at G.
   a) How do you think the angles at G are related to the angles at H? Explain.
   b) Click the button: *Slide AB to CD*.
   c) Were you correct in part a? Explain.

4. Drag points E and F to change the angles that EF makes with AB and CD.
   a) What do you notice?
   b) Why are the angles related this way?

Corresponding angles

5. Click the tab for page 2.

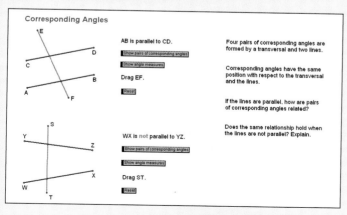

Technology: Using *The Geometer's Sketchpad* to Investigate Angles Involving Parallel Lines

6. Four pairs of corresponding angles are formed by a transversal and two lines.
   a) If the lines are parallel, how are pairs of corresponding angles related?
   b) Does the same relationship hold when the lines are not parallel? Explain.

7. Click the tab for page 3.

Alternate angles

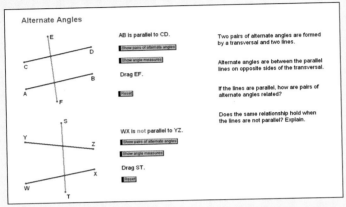

8. Two pairs of alternate angles are formed by a transversal and two lines.
   a) If the lines are parallel, how are pairs of alternate angles related?
   b) Does the same relationship hold when the lines are not parallel? Explain.

9. Click the tab for page 4.

Interior angles

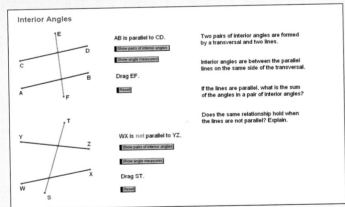

10. Two pairs of interior angles are formed by a transversal and two lines.
    a) If the lines are parallel, what is the sum of the angles in a pair of interior angles?
    b) Does the same relationship hold when the lines are not parallel? Explain.

3.3 Angles Involving Parallel Lines

Jon is learning to sail. He uses parallel rulers to map a course 225° from his current position.

He places one edge of a ruler to align with 225° on a compass.

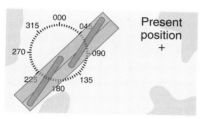

He swings the other edge of the ruler to his current position and draws the course line.

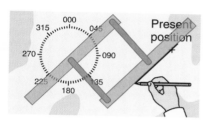

Investigate — Angles Involving Parallel Lines

You will need a protractor.

➢ Draw diagrams like this.
   Use both edges of a ruler to draw the parallel lines in the second diagram.

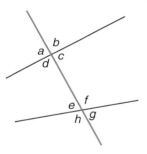 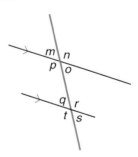

➢ Measure the angles.

➢ How are the angles related?

➢ Which angles have a sum of 180°? How do you know?

Reflect

➢ When a line intersects two parallel lines, how can you remember which pairs of angles are equal?

➢ Does each diagram have the same number of pairs of equal angles? Explain.

3.3 Angles Involving Parallel Lines   **87**

Connect the Ideas

A transversal is a line that intersects two or more other lines.

When a **transversal** intersects two lines, four pairs of opposite angles are formed. The angles in each pair are equal. When the lines are parallel, angles in other pairs are also equal. We can use tracing paper to show these relationships.

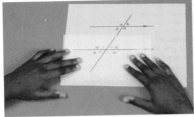

Corresponding angles form an F pattern.

Corresponding angles are equal. They have the same position with respect to the transversal and the parallel lines.

Alternate angles form a Z pattern.

Alternate angles are equal. They are between the parallel lines on opposite sides of the transversal.

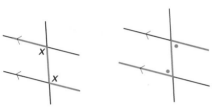

Interior angles form a C pattern.

Interior angles have a sum of 180°. They are between the parallel lines on the same side of the transversal.

We can use these relationships to determine the measures of other angles when one angle measure is known.

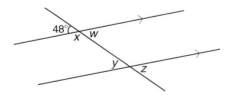

> w and 48° are measures of opposite angles. Opposite angles are equal, so $w = 48°$.

> w and x form a straight angle. The sum of their measures is 180°.
> $w = 48°$, so $x = 180° - 48°$
> $\qquad = 132°$

> w and y are alternate angles. They are equal.
> $w = 48°$, so $y = 48°$

> w and z are corresponding angles. They are equal.
> $w = 48°$, so $z = 48°$

CHAPTER 3: Relationships in Geometry

Practice

1. In each diagram, which other angle measures can you find? Determine these measures. Which relationships did you use?

a)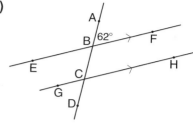

b)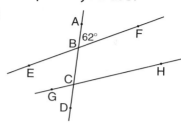

> **Example**
>
> Determine the values of x and y.
>
> **Solution** Since x and 116° are the measures of interior angles between parallel lines, their sum is 180°; that is, $x + 116° = 180°$
> So, $x = 180° - 116°$
> $= 64°$
> Similarly, the measures of y and 134° have a sum of 180°; that is,
> $y + 134° = 180°$
> So, $y = 180° - 134°$
> $= 46°$

2. Determine the angle measure indicated by each letter. Justify your answers.

a)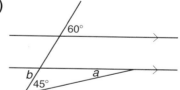

b)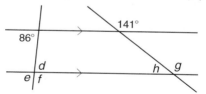

c)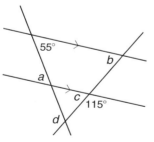

3. **Assessment Focus** Petra says these relationships are always true when two parallel lines are intersected by a transversal.
   - All the acute angles are equal.
   - All the obtuse angles are equal.
   - The sum of any acute angle and any obtuse angle is 180°.

   Is she correct?
   Explain your thinking.
   Include a diagram in your explanation.

   Which relationships can you use?

4. The diagram shows three sections of a steam pipe.
   The top pipe and bottom pipe are parallel.
   What is the angle between sections BC and CD?
   Show your work.

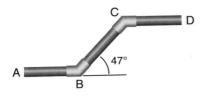

5. Alison is building a section of fence.
   To make sure the boards are parallel,
   she measures the angles indicated.
   Are the boards parallel?
   Justify your answer.

6. An ironing board is designed so the board is always
   parallel to the floor as it is raised and lowered.
   In the diagram, the two triangles are isosceles and
   ∠AEB is 88°.
   Determine all the angles in △ECD.

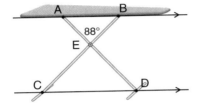

7. a) List what you know about the interior and exterior
      angles of triangles.

   Which tools could you use to help you?

   b) List what you know about angles involving parallel lines
      cut by a transversal.
   c) Use the properties in parts a and b.
      Create a design, pattern, or picture.
      Colour your design.
      Explain how it illustrates the properties.

8. **Take It Further** Use a ruler to draw a diagram like the
   one at the right.
   Two transversals intersect at B.
   a) Measure to determine the values of a, b, and c.
      How are the measures related?
   b) Draw another diagram like this, so that a, b, and c have
      different values. Are a, b, and c related in the same way
      as in part a? Explain.
   c) Explain why the relationship is always true.

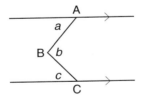

In Your Own Words

Research to find how geometry is used in real life.
You could look at occupations, sports, or architecture.
Use words, pictures, or numbers to explain what you found out.

90 CHAPTER 3: Relationships in Geometry

Polygon Pieces

The two puzzles on this page involve these 6 polygons.

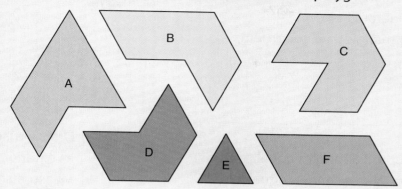

Puzzle 1

The sum of the interior angles is the same for two of the polygons. Which polygons are they?

How can you tell without measuring any of the angles and without calculating the sum of the interior angles of any of the polygons?

Puzzle 2

Your teacher will give you a copy of the 6 polygons. There is also a large regular hexagon on the copy.

Try this puzzle after you have completed Section 3.5.

- Glue your copy to a piece of cardboard. Then cut out the 6 polygons.

- Arrange the 6 polygons to fit together on the regular hexagon. Record your solution to the puzzle on triangular dot paper.

- The puzzle can be solved in different ways. Compare your solution with those of other students. Record as many solutions as you can on triangular dot paper.

Mid-Chapter Review

3.1

1. Your teacher will give you a copy of the diagram below.
The 3 triangles are congruent.

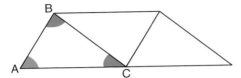

Colour red all angles equal to ∠A.
Colour blue all angles equal to ∠B.
Colour green all angles equal to ∠C.
Use the diagram to explain why the sum of the angles in a triangle is 180°.

2. Triangles ABC and ACD are braces for the "Diner" sign.
What are the values of a and b?

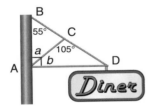

3.2

3. Use your diagram from question 1. Explain why the exterior angle of a triangle is equal to the sum of the two opposite interior angles.

4. Determine the angle measure indicated by each letter.
a)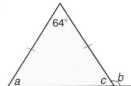

b)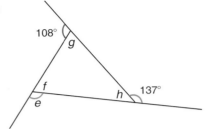

5. a) Draw a triangle and its exterior angles.
b) Measure and record enough angles so that someone could determine the remaining angles.
c) Trade diagrams with a classmate. Solve your classmate's problem.

3.3

6. Here is the flag of Antigua.

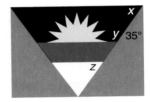

Determine the angle measure indicated by each letter.
What assumptions are you making?

7. a) Write all the relationships you can for the angles in this diagram.

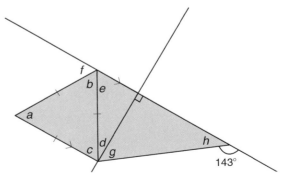

b) Use these relationships to determine the angle measure indicated by each letter.

92   CHAPTER 3: Relationships in Geometry

Using *The Geometer's Sketchpad* to Investigate Angles Related to Quadrilaterals

Part 1: The Relationship among the Interior Angles

1. Open the file *QuadAngles.gsp*. Click the tab for page 1.

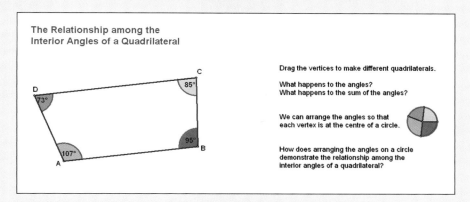

2. Drag vertices A, B, C, and D to create 5 different quadrilaterals.
   a) Copy this table. For each quadrilateral, record the measures of ∠A, ∠B, ∠C, and ∠D.

∠A	∠B	∠C	∠D	∠A + ∠B + ∠C + ∠D

   b) Calculate the sum of each set of angles. Record each sum in the table.
   c) How do the interior angles of a quadrilateral appear to be related? Explain.

3. We can arrange ∠A, ∠B, ∠C, and ∠D so the vertex of each angle is at the centre of a circle.
   a) Drag vertices A, B, C, and D to create different quadrilaterals. Observe how the angles change on the circle.
   b) How does arranging the angles on a circle support your answer to question 2c?

Technology: Using *The Geometer's Sketchpad* to Investigate Angles Related to Quadrilaterals

Part 2: The Relationship among the Exterior Angles

1. Use the file *QuadAngles.gsp*. Click the tab for page 2.

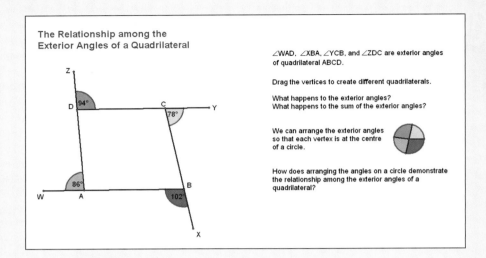

2. Drag vertices A, B, C, and D to create 5 different quadrilaterals.
   a) Copy this table. For each quadrilateral, record the measures of the exterior angles.

∠WAD	∠XBA	∠YCB	∠ZDC	∠WAD + ∠XBA + ∠YCB + ∠ZDC

   b) Calculate the sum of each set of angles. Record each sum in the table.
   c) How do the exterior angles of a quadrilateral appear to be related? Explain.

3. We can arrange the exterior angles so each vertex is at the centre of a circle.
   a) Drag vertices A, B, C, and D to create different quadrilaterals. Observe how the angles change on the circle.
   b) How does arranging the angles on a circle support your answer to question 2c?

3.4 Interior and Exterior Angles of Quadrilaterals

Investigate   **Angle Relationships in Quadrilaterals**

Which tools could you use to help you?

➤ Draw a quadrilateral.
Measure the interior angles.
Calculate their sum.

➤ Extend each side of the quadrilateral to draw 4 exterior angles, one at each vertex.
Measure the exterior angles.
Calculate their sum.

➤ Repeat the steps above for 2 different quadrilaterals.
What is the sum of the interior angles of a quadrilateral?
What is the sum of the exterior angles of a quadrilateral?

➤ What is the relationship between the interior angle and the exterior angle at any vertex of a quadrilateral?

Reflect

➤ Did your results depend on the type of quadrilateral drawn? Explain.

➤ How can you use what you know about the sum of the angles in a triangle to determine the sum of the angles in a quadrilateral?

➤ How can you use the sum of the interior angles of a quadrilateral to determine the sum of the exterior angles?

Connect the Ideas

Interior angles of a quadrilateral

Any quadrilateral can be divided into 2 triangles.

 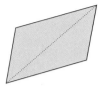 

The sum of the angles in each triangle is 180°.
The sum of the angles in 2 triangles is: $2 \times 180° = 360°$

So, the sum of the interior angles of any quadrilateral is 360°.
That is, $a + b + c + d = 360°$

Exterior angles of a quadrilateral

At each vertex, an interior angle and an exterior angle form a straight angle.
The sum of their measures is 180°.

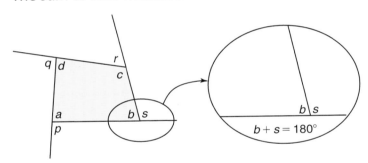

A quadrilateral has 4 vertices.
So, the sum of the interior and exterior angles is:
$4 \times 180° = 720°$
The sum of the interior angles is 360°.
So, the sum of the exterior angles is:
$720° - 360° = 360°$

The sum of the exterior angles is 360°.
That is, $p + q + r + s = 360°$

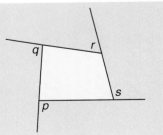

Practice

1. Draw then cut out a large quadrilateral.
   Tear off the corners.
   How can you arrange the corners to show that the
   interior angles of a quadrilateral have a sum of 360°?
   Sketch what you did.

   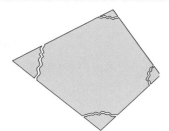

2. Determine the angle measure indicated by each letter.
   a)   b) 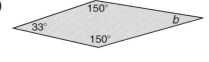   c)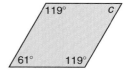

3. Draw 2 congruent parallelograms. Cut them out.
   a) Cut out and arrange the pieces of one parallelogram as shown.

   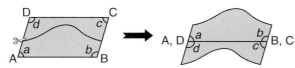

   Which tools could you use to help you?

   Cut out and arrange the pieces of the second parallelogram as shown.

   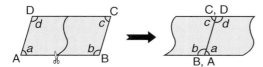

   How are two angles that share a common side of a parallelogram related?
   b) Use what you know about angles involving parallel lines
      to explain the relationship in part a.

Example

One angle of a parallelogram is 60°.
Determine the measures of the
other angles.

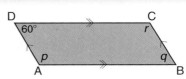

Solution

Since p and 60° are the measures of interior angles between parallel
sides DC and AB, their sum is 180°; that is, $p + 60° = 180°$
So, $p = 180° - 60°$
     $= 120°$
Similarly, p and q are the measures of interior angles between
parallel sides AD and BC.
So, $q + 120° = 180°$, then $q = 60°$

3.4 Interior and Exterior Angles of Quadrilaterals **97**

q and *r* are the measures of interior angles between parallel sides CD and BA.
So, 60° + *r* = 180°, then *r* = 120°

4. Determine the angle measure indicated by each letter. Which relationships did you use each time?
   a)
   b)
   c)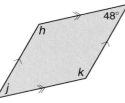

5. Draw then cut out a parallelogram. Cut out and arrange the pieces of the parallelogram as shown. What does this tell you about opposite angles in a parallelogram?
   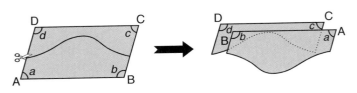

6. A kite has pairs of adjacent sides equal.

   a) Draw a kite. Determine how the angles are related. Explain how you did this.
   b) Compare your answer to part a with that of a classmate. Work together to explain why the relationships in part a are always true.

7. **Assessment Focus** Determine the angle measure indicated by each letter. Explain how you know.
   a)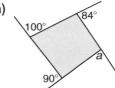
   b)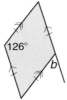
   c)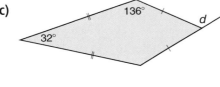

8. **Take It Further** Quadrilateral ROSE is an isosceles trapezoid. Sides OS and RE are parallel. Sides OR and SE are equal.
   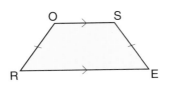
   a) Find as many relationships as you can among the interior angles of ROSE.
   b) Choose a tool and use it to demonstrate the relationships in part a.

In Your Own Words

What do you know about the interior and exterior angles of quadrilaterals? Use diagrams to explain.

98 CHAPTER 3: Relationships in Geometry

3.5 Interior and Exterior Angles of Polygons

A garden has the shape of a pentagon.
It is surrounded by straight walking paths.
James starts walking around the garden.

At each fork, he turns left.
When James returns to his starting point,
through how many degrees has he turned?

Investigate — The Sum of the Exterior Angles of a Polygon

Which tools could you use to help you?

➤ Work in a group.
You will need a large open space.
Your group will be assigned one of the polygons below.
Use masking tape to create it.

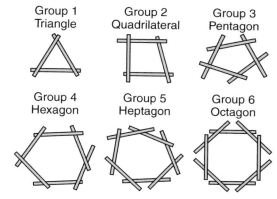

Group 1 Triangle
Group 2 Quadrilateral
Group 3 Pentagon
Group 4 Hexagon
Group 5 Heptagon
Group 6 Octagon

➤ Have a member of the group walk around the polygon until she arrives back where she started from, and faces the way she began. Determine the total angle she turned through.

➤ Repeat for each of the other polygons.
What do you notice?

Reflect

➤ Do the results depend on the number of sides in the polygon? Explain.

➤ What do these results suggest about the sum of the exterior angles of any polygon? Explain.

Connect the Ideas

Interior angles of a polygon

We can determine the sum of the interior angles of any polygon by dividing it into triangles.

Polygon	Number of sides	Number of triangles	Angle sum
Triangle	3	1	1(180°) = 180°
Quadrilateral	4	2	2(180°) = 360°
Pentagon	5	3	3(180°) = 540°
Hexagon	6	4	4(180°) = 720°

In each polygon, the number of triangles is always 2 less than the number of sides. So, the angle sum is 180° multiplied by 2 less than the number of sides.
We can write this as an equation.

$$S = (n - 2) \times 180°$$
S is the sum of the interior angles of a polygon with n sides.

A polygon with 12 sides can be divided into 10 triangles.

We can use this formula to determine the sum of the interior angles of a polygon with 12 sides.
$S = (n - 2) \times 180°$
Substitute: $n = 12$
$S = (12 - 2) \times 180°$
$ = 10 \times 180°$
$ = 1800°$
The sum of the interior angles of a 12-sided polygon is 1800°.

Exterior angles of a polygon

At each vertex, an interior angle and an exterior angle have a sum of 180°.
For 12 vertices, the sum of the interior and exterior angles is: $12 \times 180° = 2160°$
The sum of the interior angles is 1800°.

So, the sum of the exterior angles is: $2160° - 1800° = 360°$

We can reason this way for any polygon.

The sum of the exterior angles of any polygon is 360°.

Practice

1. Use a ruler to draw a large pentagon and its exterior angles as shown. Cut out the exterior angles.
   How can you arrange them to show that the exterior angles of a pentagon have a sum of 360°?
   Sketch what you did.

2. What is the sum of the interior angles of a polygon with each number of sides?
   a) 7 sides
   b) 10 sides
   c) 24 sides

3. Determine the angle measure indicated by x.
   a)
   b)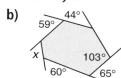

Example

A regular polygon has all sides equal and all angles equal.
What is the measure of each interior angle in a regular hexagon?

Solution A hexagon has 6 sides.
Use the sum of the interior angles:
$S = (n - 2) \times 180°$
Substitute: $n = 6$
$S = (6 - 2) \times 180°$
$= 4 \times 180°$
$= 720°$

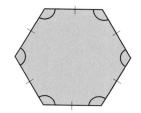

Since the angles are equal, divide by 6 to determine the measure of each angle: $\frac{720°}{6} = 120°$

Each interior angle in a regular hexagon is 120°.

4. Determine the measure of one interior angle in a regular polygon with each number of sides.
   a) 5 sides
   b) 8 sides
   c) 10 sides
   d) 20 sides

5. Determine the measure of one exterior angle for each regular polygon in question 4.

6. A Canadian $1 coin has the shape of a regular polygon.
   a) How many sides does it have?
   b) What is the measure of each interior angle?
      Show your work.

3.5 Interior and Exterior Angles of Polygons

7. **Assessment Focus**
   a) List all that you know about the interior angles and exterior angles of polygons.
   b) Use the properties in part a.
   Create a design, pattern, or picture.
   Colour your design to illustrate the properties.
   Explain your colour scheme.

   Use *The Geometer's Sketchpad* if available.

8. A concave polygon has an interior angle that is a reflex angle. Here are two examples.
   a) Draw a concave quadrilateral.
   b) Draw a concave pentagon.
   c) Measure the angles.
   Determine the sum of the angles in each figure.
   d) Do concave polygons obey this relationship:
   Interior angle sum = $(n - 2) \times 180°$?
   Justify your answer.

   A reflex angle is between 180° and 360°.

   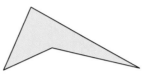

9. Find a soccer ball or use this picture.
   a) Which different polygons do you see on the ball?
   b) What is the measure of one interior angle of a white polygon?
   c) What is measure of one interior angle of a black polygon?
   d) What is the sum of the angles at the red dot?
   What do you notice?

10. **Take It Further** You can divide a hexagon into triangles by choosing a point inside the hexagon and joining all vertices to that point.
    a) What is the sum of the angles in all the triangles?
    b) What would have to be subtracted from the sum in part a to get the sum of the interior angles of the hexagon? Explain.
    c) Repeat parts a and b for other polygons.
    Write an equation for the sum of the interior angles in a polygon with *n* sides.

    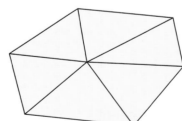

In Your Own Words

How can you find the sum of the angles in any polygon? When can you use this sum to determine the measures of the angles in the polygon? Include sketches in your answer.

Using *The Geometer's Sketchpad* to Investigate Angles Related to Regular Polygons

Part 1: Relationships among the Interior Angles and the Exterior Angles of a Regular Polygon

1. Open the file *PolyAngles.gsp*. Click the tab for page 1.

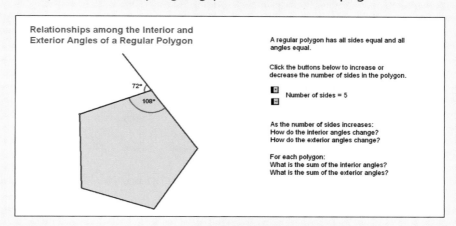

2. A regular polygon has all sides equal and all angles equal.
   a) Start with a regular polygon with 3 sides.
      This is an equilateral triangle.
      Copy the table.
      Complete the first 3 columns for an equilateral triangle.

Number of sides in the regular polygon	Measure of one interior angle	Measure of one exterior angle	Sum of interior angles	Sum of exterior angles

   b) Use the + button to create a regular polygon with 4 sides.
      Complete the first 3 columns of the table for this polygon.
   c) Repeat part b for 4 different regular polygons.
      As the number of sides increases:
      How do the interior angles change?
      How do the exterior angles change?

3. Complete the last two columns of the table for each polygon from question 2.
   As the number of sides increases:
   How does the sum of the interior angles change?
   What do you notice about the sum of the exterior angles?

Technology: Using *The Geometer's Sketchpad* to Investigate Angles Related to Regular Polygons

Part 2: The Relationship among the Exterior Angles of a Regular Polygon

1. Use the file *PolyAngles.gsp*. Click the tab for page 2.

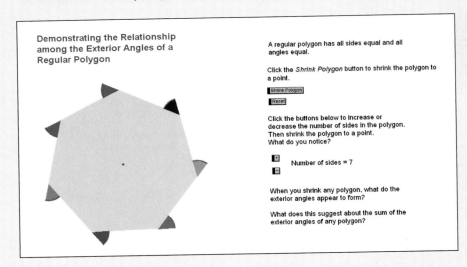

2. In Part 1, question 3, you discovered a relationship among the exterior angles of a regular polygon.
   a) Start with a triangle.
      Click the *Shrink Polygon* button to shrink the triangle to a point.
      What do the exterior angles appear to form?
      Click the *Reset* button to return the polygon to its original size.
   b) Click the + button to create a polygon with 4 sides.
      Click the *Shrink Polygon* button to shrink the polygon to a point.
      What do the exterior angles appear to form?
      Click the *Reset* button to return the polygon to its original size.
   c) Repeat the steps in part b for 4 different regular polygons.

3. a) When you shrink any regular polygon, what do the exterior angles appear to form?
   b) What does this suggest about the sum of the exterior angles of a regular polygon?

Chapter Review

What Do I Need to Know?

Any Triangle

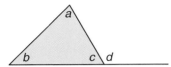

Sum of the interior angles is 180°.
$a + b + c = 180°$
Each exterior angle is the sum of the two opposite interior angles.
$d = a + b$

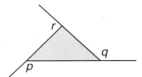

Sum of the exterior angles is 360°.
$p + q + r = 360°$

Isosceles Triangle

The angles opposite the equal sides are equal.

Equilateral Triangle

Each angle is 60°.

Parallel Lines

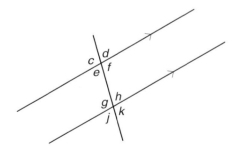

Corresponding angles are equal.
$c = g \quad e = j \quad d = h \quad f = k$
Alternate angles are equal.
$e = h \quad f = g$
Interior angles have a sum of 180°.
$e + g = 180° \quad f + h = 180°$

Any Quadrilateral

Sum of the interior angles is 360°.
$a + b + c + d = 360°$

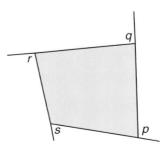

Sum of the exterior angles is 360°.
$p + q + r + s = 360°$

Chapter Review **105**

Parallelogram

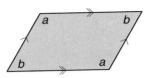

Opposite angles are equal.
Also, $a + b = 180°$

Any Polygon

If there are n sides, the sum of the interior angles is: $(n - 2) \times 180°$
The sum of the exterior angles is 360°, regardless of the number of sides.

What Should I Be Able to Do?

3.1   1. Explain how you know that the sum of the angles in a triangle is 180°.

2. Can a triangle have 2 obtuse angles? Explain.

3.1
3.2   3. Determine the angle measure indicated by each letter.
   a)

   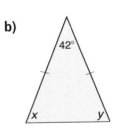

   b)

   c)

   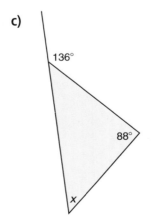

3.1
3.3   4. Determine the angle measure indicated by each letter.

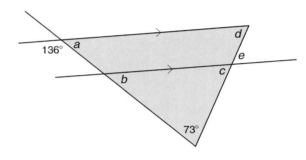

5. Are these lines parallel? Justify your answer.
   a)

   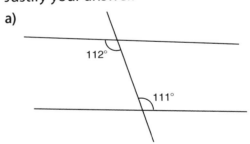

   b)

   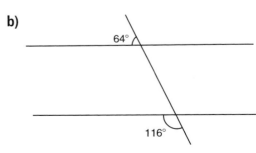

6. The top of this TV table is parallel to the ground. The triangles are isosceles. Determine the measures of the labelled angles.

3.4 | 7. Determine the angle measure indicated by each letter. Justify your answer.

a)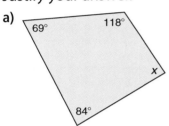

b)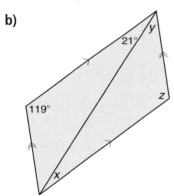

c)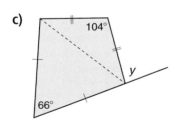

d)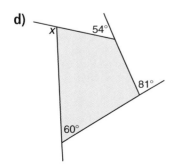

3.5 | 8. a) Use this diagram to determine the sum of the interior angles in an octagon.

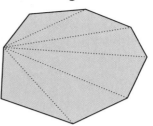

b) A regular octagon has 8 equal sides and 8 equal angles. Determine each measure:
  i) an interior angle
  ii) an exterior angle

9. Determine the sum of the interior angles of a polygon with each number of sides. Try to do this two different ways.
a) 6 sides   b) 12 sides   c) 18 sides

10. a) What is the sum of the interior angles of a pentagon? Show your work.
b) What does the following sequence of pictures demonstrate about the sum of the exterior angles of a pentagon?

c) Does the result of part a agree with your answer in part b? Explain.

11. A regular polygon has 100 sides. Determine the measure of one interior angle and one exterior angle.

Chapter Review **107**

Practice Test

Multiple Choice: Choose the correct answer for questions 1 and 2.

1. What are the values of a and b?
   - A. $a = 85°, b = 85°$
   - B. $a = 95°, b = 85°$
   - C. $a = 85°, b = 95°$
   - D. $a = 95°, b = 95°$

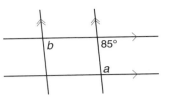

2. Determine the value of x.
   - A. 32°
   - B. 52°
   - C. 58°
   - D. 64°

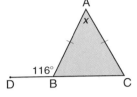

Show all your work for questions 3 to 6.

3. **Communication** Suppose you had a cutout of this quadrilateral. Explain how to demonstrate that the sum of its interior angles is 360°. How many different ways could you do this? Explain.

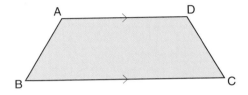

4. **Knowledge and Understanding** Determine the angle measure indicated by each letter.

   a)

   b)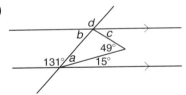

   c)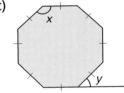

5. **Application** On a baseball diamond, home plate has the shape of a pentagon. The pentagon has 3 right angles. The other 2 other angles are equal.

   a) What is the sum of the angles of a pentagon?
   b) What is the measure of each equal angle? How do you know?

   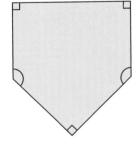

6. **Thinking** Use this diagram. Explain why $\angle PRQ = \angle RST$.

   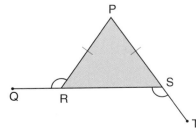

108 CHAPTER 3: Relationships in Geometry

4 Proportional Reasoning

What You'll Learn

To use different proportional reasoning strategies to solve problems involving ratios, rates, and percents

And Why

Situations involving ratios, rates, and percents occur frequently in daily life, such as calculating sales tax or getting the best deal when you shop.

Key Words

- ratio
- equivalent ratios
- proportion
- rate
- unit rate
- percent

- Organizing a Banquet

Reading and Writing in Math

Decoding Word Problems

A word problem is a math question in a real-world context.

Being able to read and understand word problems should help you relate math to life outside the classroom.

Here is a word problem:
> Joseph is buying a carpet for his living room.
> It is a rectangular room that measures 4 m by 5 m.
> The carpet costs $11.99/m^2.
> How much will Joseph spend on carpet?

Many word problems have three parts:

1. The context or set-up — Joseph is buying a carpet for his living room.

2. The math information — It is a rectangular room that measures 4 m by 5 m. The carpet costs $11.99/m^2.

3. The question — How much will Joseph spend on carpet?

Some word problems use key words to tell you what to do. Here are some common key words and phrases:

> Calculate   Estimate   Show your work
> Compare   Explain   Simplify
> Describe   Justify   Solve
> Determine   List

Work with a partner. Look through the text.
Find an example where each key word is used.
Choose one key word. Explain what it means.

CHAPTER 4: Proportional Reasoning

4.1 Equivalent Ratios

A day-care centre must maintain certain ratios of staff to children. These ratios depend on the age of the child.

For 3- to 5-year-olds, the required ratio of staff to children is 1 : 10.

Investigate　Comparing Ratios

Work with a partner.
You will need a ruler or metre stick and 10 copies of the same textbook.

➢ Measure the height of 1 textbook.
　Record it in a table.

Number of books	Height of pile (cm)	Number of books : height

➢ Pile the textbooks, one at a time.
　Record the height of the pile each time you add a book.
　Stop when your pile reaches 10 textbooks.

➢ Write the ratio of each number of books to the height.

➢ What patterns do you see in your table?

Reflect

Compare your results with those of classmates who used a different textbook.

➢ How are the patterns the same?

➢ How are they different?

4.1 Equivalent Ratios

Connect the Ideas

Recall that a **ratio** is a comparison of two quantities.
For example, suppose the ratio of students to computers in a school is 5:1.
This means there are 5 students for every 1 computer.

Equivalent ratios make the same comparison.
Suppose a class of 30 students has 6 computers.
The ratio of students to computers is 30:6.
This ratio is equivalent to 5:1, because it has the same meaning—there are 5 times as many students as computers.

Here are two ways to determine equivalent ratios.

➢ Multiply each term in a ratio by the same number.
$$8:10 = (3 \times 8) : (3 \times 10)$$
$$= 24:30$$
8:10 and 24:30 are equivalent ratios.

➢ Divide each term in a ratio by a common factor.
$$8:10 = \frac{8}{2} : \frac{10}{2}$$
$$= 4:5$$
8:10 and 4:5 are equivalent ratios.

Practice

1. Determine an equivalent ratio by multiplying.
   a) 2:3
   b) 4:3
   c) 4:5
   d) 6:5

2. Determine an equivalent ratio by dividing.
   a) 30:50
   b) 21:7
   c) 15:12
   d) 18:36

3. Determine two equivalent ratios for each ratio.
   a) 2:5
   b) 125:500
   c) 10:1
   d) 1500:1000

4. The ratio of length to width for this rectangle is 8:6.
   a) Use grid paper. Draw a smaller rectangle whose dimensions have the same ratio.
   b) Draw a larger rectangle whose dimensions have the same ratio.
   c) Amir draws a rectangle with length 26 cm and width 18 cm. Is the ratio of the dimensions of Amir's rectangle equivalent to the ratio of the dimensions of your rectangles? How do you know?

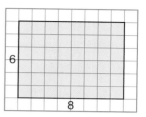

The maximum height allowed for a wheelchair ramp is 76 cm.

5. An architect draws a plan for a wheelchair ramp. On the plan, the ramp is 2 cm high and 24 cm long. What might the dimensions of the actual ramp be? How did you use equivalent ratios to find out?

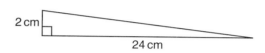

We can use equivalent ratios to compare ratios.

> **Example**
>
> June and Megan make grape punch for a party.
> June uses grape concentrate and ginger ale in the ratio 3:2.
> Megan uses grape concentrate and ginger ale in the ratio 8:5.
> Which punch has the stronger grape taste? Explain.
>
> **Solution**
>
> The recipe that uses more concentrate for the same volume of ginger ale has the stronger grape taste.
>
> The volume of ginger ale is the second term of each ratio.
> So, write 3:2 and 8:5 with the same second term.
> The least common multiple of 2 and 5 is 10.
> So, write both 3:2 and 8:5 with the second term 10.
>
> **June's punch**
> 3:2 = (3 × 5):(2 × 5)
>   = 15:10
>
> **Megan's punch**
> 8:5 = (8 × 2):(5 × 2)
>   = 16:10
>
> The volume of concentrate is the first term of each ratio.
> Megan's punch uses more concentrate for the same volume of ginger ale.
> It has the stronger grape taste.

6. Refer to the *Guided Example,* page 113.
   a) Write equivalent ratios for 3:2 and 8:5 with the same first term.
   b) How do the ratios in part a help you determine which punch has the stronger grape taste?

7. Compare the strawberry punch in each pair of pitchers. Which has the stronger strawberry taste? Explain.
   a)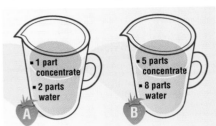
   b)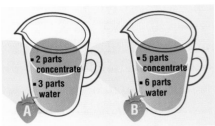

8. **Assessment Focus** Beakers of blue and clear liquids are combined.
   a) Predict which set you think will produce the bluer liquid. Justify your answer.

   b) Which set does produce the bluer liquid? Explain.
   c) Sketch a different set of beakers that would produce the same shade of blue as set A.

9. A loft measures 10 m by 16 m.
   A scale drawing of the loft is to fit on a 21-cm by 28-cm sheet of paper.
   What could the dimensions of the scale drawing be? How did you find out?

10. **Take It Further** A CD jewel case measures 12.5 cm by 14.0 cm by 1.0 cm.
    A model of the case measures 5.0 cm by 5.6 cm by 0.5 cm.
    One dimension on the model is incorrect.
    a) Which dimension is incorrect?
    b) What should the dimension be?

In Your Own Words

What are the equivalent ratios?
Explain two ways to determine a ratio that is equivalent to a given ratio.
Include examples in your explanation.

4.2 Ratio and Proportion

A nurse uses equivalent ratios to calculate drug dosages.
A drug comes in 150-mg tablets.
The dosage ordered is 375 mg.
The nurse needs to calculate how many tablets are required.

Investigate: Using Proportional Reasoning to Solve a Problem

Use counters.

A hospital gift shop sells 4 magazines for every 3 books.

➤ Suppose the shop sold 18 books. How many magazines did it sell?

➤ Suppose the shop sold 12 magazines. How many books did it sell?

➤ One Friday, the shop sold 35 books and magazines altogether. How many of each kind did it sell?

Record your work.

Reflect

Compare your recording strategies with those of a classmate.

➤ How could you solve the problems without using counters?

➤ How can you use equivalent ratios to solve the problems?

Connect the Ideas

Hamadi is a pediatric nurse. He uses a table like the one below to give the correct dose of a pain reliever.

Approximate body mass (kg)	5	10	15
Dose (mg)	60	120	180

×2 (from 5 to 10); ×3 (from 60 to 180)

When the body mass doubles, the dose doubles.

When the body mass triples, the dose triples.

We say that the drug dose is *proportional* to the body mass.
5 : 60, 10 : 120, and 15 : 180 are equivalent ratios.

We can determine the drug dose for a body mass of 35 kg.
Let d milligrams represent this dose.
We need a ratio equivalent to 5 : 60, with the first term 35.
That is, 5 : 60 = 35 : d

Here are two ways to determine the value of d.

> In proportional situations, the quantities involved are related by multiplication or division.

> The statement 5 : 60 = 35 : d is a **proportion**. A proportion is a statement that two ratios are equal.

> Use a table.

Look for a multiplication relationship *between* ratios

5 : 60 = 35 : d

Think: What do we multiply 5 by to get 35? Multiply 60 by the same number.

×7:
5	60
35	d
 :×7

So, 60 × 7 = d
That is, d = 420

Look for a multiplication relationship *within* ratios

5 : 60 = 35 : d

Think: What do we multiply 5 by to get 60? Multiply 35 by the same number.

×12:
5	60
35	d
 :×12

So, 35 × 12 = d
That is, d = 420

A dose of 420 mg is needed for a body mass of 35 kg.

Practice

1. Describe two ways the numbers in each proportion are related.
   a) $5:20 = 125:500$ b) $10:1 = 120:12$ c) $75:25 = 300:100$ d) $1:3 = 16:48$

2. Multiply between ratios to determine each value of n.
   a) $2:5 = 8:n$ b) $2:n = 6:9$ c) $n:5 = 12:20$ d) $8:n = 4:15$

3. Multiply within ratios to determine each value of z.
   a) $4:8 = 3:z$ b) $5:z = 6:18$ c) $z:14 = 10:20$ d) $3:21 = z:56$

4. A portable music player with 4 GB of memory stores about 1000 songs.
   A music player with 60 GB of memory stores about 15 000 songs.
   Is the number of songs proportional to the amount of memory?
   Explain your reasoning.

5. To make green paint, 3 parts yellow paint are mixed with 2 parts blue paint.
   Janis has 12 L of blue paint. How much yellow paint does she need? Explain.

6. Ali earned $80 working 10 h.
   How long would it take him to earn $200? Explain how you found your answer.

7. A recipe that serves 4 people uses 3 potatoes.
   How many potatoes are needed to serve 20 people?

Need Help?
Read Connect the Ideas.

Sometimes, it is helpful to simplify one of the ratios in a proportion.

Example Determine the value of c.
$c:20 = 18:15$

Solution $c:20 = 18:15$
We cannot immediately identify how the terms are related.

$c:20 = 18:15$

Use mental math to determine an equivalent ratio for $18:15$.
Divide each term by 3.
$18:15 = \frac{18}{3}:\frac{15}{3}$
$= 6:5$
So, $c:20 = 6:5$
$\div 4$

Since $20 \div 4 = 5$, then $c \div 4 = 6$
So, $c = 24$

4.2 Ratio and Proportion

8. Determine the value of each variable.
   a) $4:10 = 18:c$
   b) $125:25 = n:6$
   c) $6:y = 9:12$
   d) $60:z = 24:6$

9. **Assessment Focus** In 1996, the Royal Canadian Mint issued the new $2 coin.
   a) A poster advertising the new coin showed a large photograph of the toonie.
      The diameter of the inner core on the poster was 51 cm. What was the outer diameter of the coin on the poster? Show your work.
   b) Did you solve the problem using a proportion? How could you solve it without using a proportion?

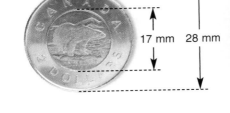

10. When a robin flies, it beats its wings about 23 times in 10 s. How many times will it beat its wings in 2 min? Explain your thinking.

11. A gear ratio is the ratio of the numbers of teeth in two connected gears.
    The gear ratio of two gears is $3:2$.
    a) The larger gear has 126 teeth.
       How many teeth does the smaller gear have?
       How did you solve the problem?
    b) Suppose the smaller gear has 126 teeth.
       How many teeth would the larger gear have? Explain.
    Show your work.

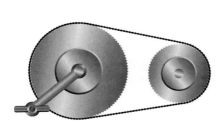

12. **Take It Further** There are 900 students enrolled in Mount Forest Secondary School.
    The ratio of girls to boys is $5:4$.
    a) How many boys and how many girls go to Mount Forest SS? Explain how you found your answer.
    b) The average class size is 27 students.
       Suppose this class is representative of all the students in the school.
       How many students in this class are girls?
       How many are boys?

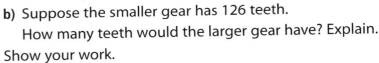

Use one of the questions in this section.
Explain how you can use a proportion to answer the question.

Using *The Geometer's Sketchpad* to Investigate Scale Drawings

Identifying Scales

1. Open *The Geometer's Sketchpad*.
   Open the sketch *Investigating Scale Drawings.gsp*.
   The sketch has scale drawings of some rooms in a house.
   The actual dimensions of each room are shown.

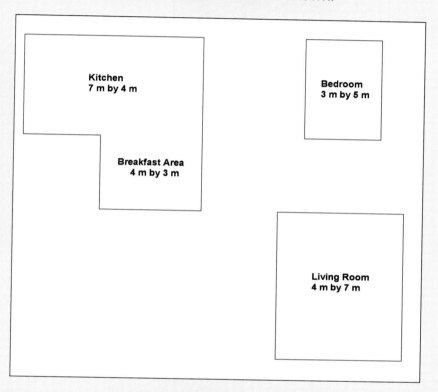

Use the instructions your teacher gives you.

2. Measure the lengths of the sides of each room.

3. Compare the actual measures on the diagram to the lengths you measured.
   a) Which rooms are drawn to scale? How do you know?
   b) What is the scale?

4. a) Which rooms are not drawn to scale? Explain.
   b) Suppose these rooms were drawn using the scale in question 3.
      What would the dimensions be?

Creating a Scale Drawing

1. Open a new sketch.

2. You will create a scale drawing of a soccer field. The actual dimensions of the field are:

   Touchline – 120 m

   Goal line – 90 m

   Penalty area – 16.5 m by 40 m

   Centre circle – radius of 9.15 m

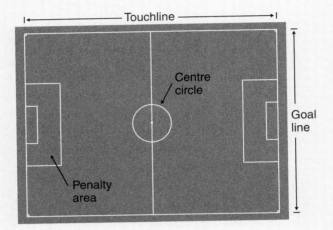

   Use the instructions your teacher gives you to construct a scale drawing.

3. Choose the scale you will use to create your drawing.

4. Follow the instructions to create the scale drawing using the scale you chose. How do you determine the length and width of your scale drawing?

5. How would the drawing change if you used a smaller scale? A larger scale?
   Construct new scale drawings to check your predictions.
   Did you use equivalent ratios? Explain.

6. Print your scale drawings.
   Trade drawings with a classmate.
   Determine the scale for each of your classmate's drawings.

7. Construct a scale drawing of an object of your choice.
   Print the drawing and explain how you constructed it.
   Include the actual dimensions of the object and the scale.

4.3 Unit Rates

In many places in Ontario, home-owners use firewood to heat their homes.
The firewood is split and stacked into units called *cords*. One cord of wood measures about 1.20 m wide by 2.40 m long by 1.20 m high.

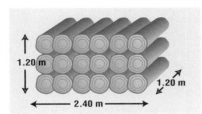

Investigate Comparing Unit Rates

You will investigate two methods in which firewood is split.

➢ Northern Lights Firewood uses a machine called a firewood processor.
  It produces 32 cords in 8 h.

➢ Duhaime's Timber Products uses 6 people with chain saws and log splitters.
  They produce 21 cords in 6 h.

Which company produces firewood faster?

Reflect

➢ What strategy did you use to determine the faster method?

➢ Compare your strategy for solving the problem with those of other students.
  If you used different strategies, explain your strategy to other students.

Connect the Ideas

In each case, the operating costs are proportional to the time.

Northern Lights Firewood
Operating costs: $620 for 8 h

Duhaime's Timber Products
Operating costs: $585 for 6 h

To determine which company has the greater operating costs, we determine the operating costs per hour.

Northern Lights Firewood's operating costs are $620 for 8 h. To determine the hourly operating costs, divide 620 by 8.
$620 \div 8 = 77.5$
The operating costs are $77.50 per hour.

$\div 8 \quad \begin{array}{c|c} 620 & 8 \\ \hline ? & 1 \end{array} \quad \div 8$

Duhaime's Timber Products' operating costs are $585 for 6 h. To determine the hourly operating costs, divide 585 by 6.
$585 \div 6 = 97.5$
The operating costs are $97.50 per hour.
So, Duhaime's Timber Products has the greater operating costs.

$\div 6 \quad \begin{array}{c|c} 585 & 6 \\ \hline ? & 1 \end{array} \quad \div 6$

A **rate** is a comparison of two quantities with different units.

Each hourly rate is called a **unit rate**, because it tells the cost for 1 h, or one *unit* of time.

Unit rates can be written in different ways:

➢ Using words: one hundred kilometres per hour

➢ Using numbers, symbols, and words: 100 km per hour

➢ Using numbers and symbols: 100 km/h

122 CHAPTER 4: Proportional Reasoning

Practice

1. Determine each unit rate.
   a) 50 goals scored in 25 games
   b) $400 earned in 40 h
   c) $6.00 for 12 oranges
   d) 770 km travelled in 7 h

2. Determine each unit rate.
   a) $5.00 for 10 CDs
   b) $2.37 for 3 kg of apples
   c) 9 kg lost in 6 weeks
   d) 22 km hiked in 5 h

3. Twelve hundred litres of water were pumped out of a flooded basement in 8 h. What was the unit rate of outflow?

4. Mr. and Mrs. Dell recently finished some renovations to their house.
   Twenty square metres of carpeting cost $875.
   Seventeen square metres of hardwood flooring cost $750.
   Which flooring is more expensive per square metre? Explain.

5. Alec wants to join one of three online music clubs.

eTunes	**Tunezilla**	**Monstersongs.com**
$21.00 for 20 songs	$41.65 for 35 songs	$54.50 for 50 songs

   Which music club is more economical? Explain.

6. A 12-pack of pop costs $2.99.
   The same pop costs $0.75 per can from a vending machine.
   Is this a fair price? Explain.

We can use unit rates to determine the best buy.

Example

Cereal comes in 3 sizes.

a) Calculate the unit price for each box.
b) Which box is the best buy? Explain.

Solution a) Since 1 g is very small, we calculate the cost of 100 g.
This is the unit price.
Box A has mass 500 g and costs $2.99.
500 g is 5 × 100 g; so, the cost of 100 g of Box A is: $\frac{\$2.99}{5} = \$0.598$

Box B has mass 650 g and costs $3.69.
650 g is 6.5 × 100 g; so, the cost of 100 g of Box B is: $\frac{\$3.69}{6.5} \doteq \$0.568$

Box C has mass 800 g and costs $4.29.
800 g is 8 × 100 g; so, the cost of 100 g of Box C is: $\frac{\$4.29}{8} \doteq \$0.536$

b) Box C has the lowest unit price. It is the best buy.

7. Which is the best buy? Show your work.

8. **Assessment Focus** Sue wants to buy tea.
The brand that she buys comes in 2 sizes.
   a) Use unit rates.
   Which box of tea bags is the better buy?
   b) How could Sue determine the better buy without using unit rates? Explain.
   c) Why might Sue not want to purchase the better buy?

9. **Take It Further** A 400-g package of Grandma's Cookies has 18 cookies and costs $2.49.
A 700-g package of Dee's Delights has 12 cookies and costs $3.99.
   a) Which unit would you use to determine the better buy?
   Explain why you chose that unit.
   b) Determine the unit price of each brand of cookie.
   c) Which brand is the better buy? Explain.

In Your Own Words

What is a unit rate?
How can you use unit rates to determine the better buy?
Use examples in your explanation.

Grid Paper Pool

Materials

Grid Paper Pool Master
1-cm grid paper

Each "pool table" is a rectangle drawn on a grid.
There is a pocket at each corner.
The figure at the right is a 3 by 2 table.

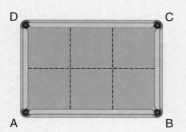

Here are the rules of the puzzle.

➤ An imaginary ball is hit from corner A. It travels across the grid, at 45° to AB.

➤ The ball bounces off each side it hits at a 45° angle.

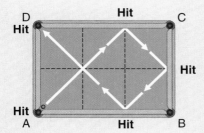

➤ The ball continues to travel until it reaches a pocket.

Your teacher will give you copies of different "pool tables."

➤ For each pool table:
 – Draw the path the ball takes.
 – Copy and complete this table.
   Write the pocket letter in the third column.

Pool table dimensions	Number of hits	Ball reaches pocket:

➤ What patterns do you see in the results?

➤ Another table measures 16 by 24.
 Predict the number of hits before the ball stops.
 Predict the pocket in which the ball stops.
 How did you make your predictions?

➤ Determine the dimensions of 2 different pool tables where the ball stops in pocket D after 7 hits.

The ball always travels along diagonals of squares.

Both where the ball starts and where it ends are counted as hits.

Puzzle: Grid Paper Pool **125**

Mid-Chapter Review

4.1

1. Which ratios in each pair are equivalent? How do you know?
   a) 12:15 and 48:60
   b) 3:2 and 25:16
   c) 3:4 and 15:18
   d) 625:125 and 180:36

2. Lemonade is made with water and concentrate.
   a) Which pitcher has the stronger lemonade? Explain how you know.

   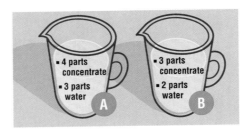
   - 4 parts concentrate
   - 3 parts water
   A
   - 3 parts concentrate
   - 2 parts water
   B

   b) Describe the contents of a different pitcher that has the same strength lemonade as Pitcher B.

4.2

3. A 7500-kg African elephant eats about 200 kg per day. A 5000-kg African elephant eats about 150 kg per day. Is the amount of food an elephant eats proportional to its mass? Explain your reasoning.

4. Determine the value of each variable.
   a) $5:3 = n:24$
   b) $4:7 = m:35$
   c) $16:4 = 12:y$
   d) $18:21 = r:14$

5. A recipe for salad dressing uses 60 mL of vinegar for 240 mL of oil. Emil has only 40 mL of vinegar. How much oil should he use?

6. In 2002, 3 out of 10 doctors in Canada were women.
   a) What is the ratio of female doctors to male doctors?
   b) There are 72 female doctors in a city. How many male doctors are there?
   c) There are 119 male doctors in another city. How many female doctors are there?
   d) What assumptions did you make in parts b and c?

4.3

7. Calculate each unit rate.
   a) Driving 45 km on 3 L of gas
   b) Typing 120 words in 4 min
   c) Earning $148 for an 8-h shift
   d) Driving 525 km in 6 h

8. Beckie and Scott rode their bicycles to school.
   Beckie rode 8 km in 15 min.
   Scott rode 12 km in 25 min.
   Who rode faster? Explain.

9. a) Determine the price for 100 g of each box of cereal.

   CEREAL $4.69 450 g A
   CEREAL $6.49 600 g B

   b) Which box is the better buy? Explain.

126 CHAPTER 4: Proportional Reasoning

4.4 Applying Proportional Reasoning

Cary wants to estimate the distance between two places. He uses a map scale and proportional reasoning.

Many different types of problems can be solved using proportional reasoning.

Investigate — Using Different Strategies to Solve a Problem

➤ Use Pattern Blocks to create this design.

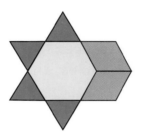

Each block in the design has a value that is proportional to its area.
The value of the yellow hexagon is $3.
What is the value of the entire design?
How do you know?

➤ Create a design that costs between $5 and $6.
Write the value of your design.
Record your design.
Explain how you determined the value.

Reflect

Compare strategies with your classmates.

➤ How could you use equivalent ratios to solve the problems?

➤ How could you use unit rates to solve the problems?

Connect the Ideas

Kirsten's neighbour used 120 patio stones to cover 8 m². Kirsten wants to build a patio using the same type of stone.

Use reasoning

Kirsten wants to pave 18 m² of her backyard.
We can use unit rates to determine the number of patio stones Kirsten needs.
An area of 8 m² requires 120 patio stones.
So, an area of 1 m² requires: $\frac{120}{8}$ stones = 15 stones
Then, an area of 18 m² requires: 18 × 15 stones = 270 stones
Kirsten needs 270 stones to pave her backyard.

Kirsten discovers that the patio stones are on sale.
She can afford to buy 330 stones.
Kirsten decides to build a larger patio.

Use a table

We can use a similar method to determine the greatest area Kirsten can cover with 330 stones.

	120	8
÷ 120	1	$\frac{8}{120}$
× 330	330	$330 \times \frac{8}{120}$

(÷ 120 and × 330 applied to both columns)

$330 \times \frac{8}{120}$ m² = 22 m²
The greatest area Kirsten can cover with 330 stones is 22 m².

Practice

1. Sean buys 8 DVDs for $120.
   Each DVD costs the same amount.
   a) How much does 1 DVD cost?
   b) How much do 13 DVDs cost?

2. Ioana works for 4 days and earns $224. At this rate:
   a) How much does she earn in 1 day?
   b) How much does she earn in 11 days?

3. Mika worked 6 shifts and earned $720.
   She earns the same amount for each shift.
   How much money would she earn in 4 shifts?

4. Zeljko purchased 4 cases of pop for $26.
   a) How many cases could he buy for $91? Explain.
   b) How could you have solved the problem a different way?

5. A car travels 96 km on 8 L of gasoline.
   a) How far will the car travel on 25 L of gasoline?
   b) How much gasoline is needed to travel 120 km?
   Show your work.

6. A machine can bind 1000 books in 8 min.
   a) How many books can the machine bind in 1 h?
   b) How long will it take to bind 7250 books? Explain.

We use proportional reasoning when we use a map to determine distances.

Example   The driving distance from Sault Ste. Marie to North Bay is about 425 km.
On a map, this distance is 8.5 cm.
On the same map, the distance from North Bay to Windsor is 13.5 cm.
What is the actual distance between these cities?

Solution   8.5 cm represents 425 km.
So, 1 cm represents: $\frac{425}{8.5}$ km = 50 km
Then, 13.5 cm represents: 13.5 × 50 km = 675 km
The actual distance from North Bay to Windsor is about 675 km.

4.4 Applying Proportional Reasoning

7. Madurodam is a miniature city in the Netherlands.
   Every object in the city is built to the same scale.
   A lamppost in Madurodam is 12 cm tall.
   It is modelled after an actual lamppost that is 3 m tall.
   a) A bridge that crosses the canal in Madurodam is 30 cm long. How long is the actual bridge that was used as the model?
   b) A clock tower is to be built in Madurodam. It is modelled after a clock tower that is 25 m tall. How tall should the miniature tower be?
   c) Explain how you solved each problem.

8. Chloë challenged her father to a basketball game.
   Each person gets a different number of points for scoring a basket.
   a) Chloë scored 12 baskets and has 36 points. How many points would she get if she scored 15 baskets?
   b) Her father scored 16 baskets and has 32 points. How many points would he get if he scored 19 baskets?
   c) Their total score at the end of the game is 95 points. How many baskets did each person score?

9. **Assessment Focus** Imran is visiting his grandmother in the United States. He can exchange $20 Canadian for $17 US.
   a) How many US dollars would Imran get for $225 Canadian?
   b) When Imran returns, he has $10 US. How much is this in Canadian dollars? What assumptions did you make?

10. **Take It Further** A recipe for pancakes calls for 2 cups flour mixed with $1\frac{1}{2}$ cups buttermilk.
    Liisa mixes 5 cups flour with $2\frac{1}{2}$ cups buttermilk.
    Will Liisa's pancake mix be too dry, too watery, or just right?
    How do you know?

In Your Own Words

You have learned several strategies to solve proportional reasoning problems. Choose one strategy. Explain when you would use it. Include an example of a problem.

4.5 Using Algebra to Solve a Proportion

Hockey sticks vary in length.
Taller players use longer sticks.
The National Hockey League only allows hockey sticks
that are shorter than 1.60 m.

Length

Investigate Solving an Equation to Solve a Problem

In 2005, Zdeno Chara was a defenceman for the Ottawa Senators.
He was the tallest player in the league.
The NHL made an exception to the stick length rule for Chara because he was so tall.
At 2.06 m, Chara used a stick that was 1.80 m long.

➤ The shortest player in the NHL has a height of 1.70 m.
Suppose the ratio of the length of his stick to his height is the same as it is for Chara.
What is the length of the stick?

➤ The Ottawa Senators sometimes give out souvenir miniature sticks at their games.
How short would a player have to be to use a stick 37 cm long?
What assumptions do you make?

Reflect

Compare strategies for solving these problems with your classmates.

➤ Did you write a proportion in each case? Explain.

➤ What other strategies could you have used to solve these problems?

➤ How could you check your answers?

Connect the Ideas

In hockey, when one team receives a penalty and plays one player short, the other team is said to be on a power play.

The Toronto Maple Leafs scored 17 power-play goals in the first 8 games of the 2005–2006 season. Suppose the team continued to score power-play goals at the same rate. We can estimate how many of these goals it might score in the 82-game season.

Let n represent the number of power-play goals scored in 82 games.
Here are two ways to determine the value of n.

In the left column, we compare ratios of goals to games.

In the right column, we compare the ratio of goals to the ratio of games.

An equation is a balance. To keep the balance, what we do to one side we must also do to the other side.

Solve a proportion using *within* ratios

17 goals … in … 8 games
n goals … in … 82 games

$17:8 = n:82$
Write each ratio in fraction form.
$$\frac{17}{8} = \frac{n}{82}$$
To isolate n, multiply each side of the equation by 82.
$$82 \times \frac{17}{8} = \frac{n}{82} \times 82$$
$$\frac{82 \times 17}{8} = \frac{82n}{82}$$
$$\frac{1394}{8} = n$$
$$n = 174.25$$

Solve a proportion using *between* ratios

17 goals … in … 8 games
n goals … in … 82 games

$n:17 = 82:8$
Write each ratio in fraction form.
$$\frac{n}{17} = \frac{82}{8}$$
To isolate n, multiply each side of the equation by 17.
$$17 \times \frac{n}{17} = \frac{82}{8} \times 17$$
$$\frac{17n}{17} = \frac{82 \times 17}{8}$$
$$n = \frac{1394}{8}$$
$$n = 174.25$$

If the Leafs continued at the same rate as for the first 8 games, they would have scored about 174 power-play goals in the 2005–2006 season.

Estimate to check reasonableness

The ratio of power-play goals to games is $17:8$.
This is approximately $16:8$, or $2:1$.
So, the number of power-play goals is about double the number of games.
Then, in 82 games, the number of power-play goals would be about: $82 \times 2 = 164$
This is close to the calculated number of 174.

132 CHAPTER 4: Proportional Reasoning

Practice

1. Solve for n.
   a) $\frac{n}{6} = \frac{3}{2}$
   b) $\frac{n}{4} = \frac{6}{8}$
   c) $\frac{n}{12} = \frac{5}{4}$
   d) $\frac{n}{25} = \frac{8}{5}$
   e) $\frac{n}{16} = \frac{5}{8}$

2. Solve for each variable.
   a) $c:9 = 21:7$
   b) $m:840 = 1:120$
   c) $y:63 = 5:3$
   d) $a:225 = 7:9$

The length of an object's shadow is proportional to the object's height.

Example

A person is 1.85 m tall.
At a certain time of day, her shadow is 0.74 m long.
At the same time of day, the shadow of a tree is 8.40 m long.
How tall is the tree?

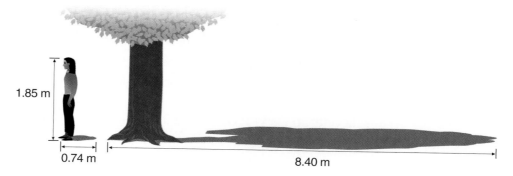

Solution

Let n metres represent the height of the tree.
The ratio of the person's height to the length of her shadow is 1.85:0.74.
The ratio of the tree's height to the length of its shadow is $n:8.40$.
Write a proportion.
$1.85:0.74 = n:8.40$
Write each ratio in fraction form.
$\frac{1.85}{0.74} = \frac{n}{8.40}$
To isolate n, multiply each side of the equation by 8.40.
$8.40 \times \frac{1.85}{0.74} = \frac{n}{8.40} \times 8.40$
$\frac{8.40 \times 1.85}{0.74} = \frac{8.40n}{8.40}$
Simplify the equation.
$\frac{15.54}{0.74} = n$
$n = 21$
The tree is 21 m tall.

3. Each shadow is measured at the same time as the shadows in the *Guided Example* on page 133.
   a) A toddler casts a shadow that is 0.33 m long.
      How tall is the toddler?
   b) A flagpole casts a shadow that is 2.96 m long.
      How tall is the pole?

4. At an automotive repair shop, 3.5 h of labour costs $311.50.
   What is the labour charge for a 5-h job?
   How could you check your answer?

5. **Assessment Focus**
   a) In 2005, the band U2 sold out two concerts in Dublin, Ireland.
      Forty-eight thousand tickets were sold in the first 15 min.
      Assume the tickets sold at a constant rate.
      How many tickets were sold in 50 min?
   b) Seventy-five thousand tickets were sold at the Stade de France in Paris, France.
      Suppose the tickets sold at the same rate as in Dublin.
      How long did it take to sell the 75 000 tickets?
   c) Explain the strategies you used to solve the problems.

6. **Take It Further** Leonardo Da Vinci believed in ideal proportions of the body.
   In his paintings, the ratio of a person's height to the floor-to-hips height is 7 : 4.
   a) A person has a floor-to-hips height of 1.16 m.
      How tall is the person?
   b) Elisha is 167 cm tall.
      What is her floor-to-hips height?
   c) Determine whether the ratio of your height to floor-to-hips height is the same as Da Vinci's ratio.
      Compare your findings with those of your classmates.

In Your Own Words

Use one of the questions in this section.
If you solved the problem using an equation, find a different way to solve it.
If you did not use an equation to solve the problem,
solve it now by using an equation.

4.6 Percent as a Ratio

Investigate
Using Percent to Compare Raises

Work with a partner.

Elena, Shane, and Gillian worked last summer at different retail stores.
This summer, all of them received raises.

Elena's wages increased from $360 to $425 per week.
Shane's wages increased from $275 to $340 per week.
Gillian's wages increased from $400 to $470 per week.

➤ Who received the greatest raise?

➤ Compare each person's raise with the previous year's wages. Who received the greatest percent increase?

Explain how you know.

Reflect

Compare strategies with your classmates for solving these problems.

➤ You represented each pay raise in two ways. Which way do you think better represents the pay raise? Explain.

Connect the Ideas

A **percent** is a ratio that compares a number to 100; for example, 30 : 100.
A percent can also be written with a symbol (%), as a fraction, or as a decimal.
For example, $30\% = \frac{30}{100} = 0.30$

An MP3 player that regularly sells for $180 is on sale for 25% off.

Determine the sale price

The sale price is 100% − 25% = 75% of the original price, or 75% of $180.
So, the sale price is in the same ratio to 180 as 75 is to 100.

Let b dollars represent the sale price.
$b : 180 = 75 : 100$
Write this proportion in fraction form.
$\frac{b}{180} = \frac{75}{100}$
$\frac{b}{180} = 0.75$

To isolate b, multiply each side of the equation by 180.
$\frac{b}{180} \times 180 = 0.75 \times 180$
$b = 135$

Notice that 75% of 180 is 0.75 × 180.

The sale price is $135.

Determine the total price

We have to pay 14% in taxes on the MP3 player.
So, the total price is 100% + 14% = 114% of the sale price, or 114% of $135.

> In Ontario, in July 2006, the provincial sales tax (PST) was 8% and the goods and services tax (GST) was 6%.

Let t dollars represent the total price.
$\frac{t}{135} = \frac{114}{100}$
$\frac{t}{135} = 1.14$

To isolate t, multiply each side of the equation by 135.
$\frac{t}{135} \times 135 = 1.14 \times 135$
$t = 153.9$

Notice that 114% of 135 is 1.14 × 135.

The total price of the MP3 player is $153.90.

136 CHAPTER 4: Proportional Reasoning

Practice

1. Write each fraction as a decimal.
   a) $\frac{7}{100}$
   b) $\frac{15}{100}$
   c) $\frac{35}{100}$
   d) $\frac{80}{100}$
   e) $\frac{120}{100}$

2. Determine each value.
   a) 10% of $365
   b) 25% of 50 kg
   c) 50% of 28 m
   d) 125% of 120 g

3. A jacket was regularly priced at $159.99.
   It was marked down by 30%. What was the sale price of the jacket?

4. During January, 50 000 new vehicles were sold in Ontario.
   About 20% of these were leased. How many vehicles were leased?

5. A video game sells for $59.99.
   a) How much is the sales tax on the game?
   b) What is the price including taxes?

6. Skis regularly sell for $350.
   They are on sale at 45% off. What is the total cost, including taxes?
   Think of a different way to solve this problem.

7. There were 288 spectators at the football game.
   75% were cheering for the home team.
   a) How many spectators were cheering for the home team? Explain.
   b) 40% of the spectators were students.
      How many spectators were adults? How do you know?

We use percents when we calculate simple interest.

> **Example**
> Emma borrows $1200 for 6 months.
> The annual interest rate is 6%.
> How much simple interest does Emma pay?
>
> **Solution** *Method 1:* Use ratios
> For one year, the ratio of simple interest to the loan
> is equal to the ratio of the interest rate to 100%.
> Let x dollars represent the simple interest.
> Then, $\frac{x}{1200} = \frac{6}{100}$
> $1200 \times \frac{x}{1200} = \frac{6}{100} \times 1200$
> $x = \frac{7200}{100}$
> $x = 72$

4.6 Percent as a Ratio

The interest for 1 year is $72, so the interest for 6 months is: $\frac{\$72}{2} = \$36$

Method 2: Use algebra
a) Use the formula: $I = Prt$
   I is the simple interest in dollars.
   The principal, P, is $1200.
   The annual interest rate, r, is 6%, or 0.06.
   Since time, t, is measured in years, write 6 months as a fraction of a year: $\frac{6}{12}$
   Substitute: $P = 1200$, $r = 0.06$, and $t = \frac{6}{12}$
   So, $I = 1200 \times 0.06 \times \frac{6}{12}$
   $= 36$
   Emma pays $36 simple interest.

8. Connor borrows $5000 for 9 months. The annual interest rate is 8%. How much simple interest does Connor pay?

9. John put $500 in a savings account for 8 months. The annual interest rate is 2%.
   a) How much simple interest does the money earn?
   b) How much money is in the account after 8 months?

10. **Assessment Focus** A credit card company charges 24% per year on outstanding balances.
    a) How much interest would be charged on an outstanding balance of $900 for 90 days?
    b) How much is owed at the end of 90 days?
    Show your work.

11. **Take It Further** Marie borrowed $3500 for 6 months. She paid $140 simple interest. What was the annual interest rate? How could you check your answer?

In Your Own Words

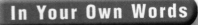

What do you find most challenging when you solve problems involving percent?
Use a question from this section to explain.
How might you overcome this difficulty?

Chapter Review

What Do I Need to Know?

A **ratio** is a comparison of two quantities.
Two ratios are **equivalent** when they can be reduced to the same ratio.
For example, both 12 : 16 and 9 : 12 reduce to 3 : 4, so they are equivalent ratios.

> Use a Frayer model to help you understand some of the key words on this page.

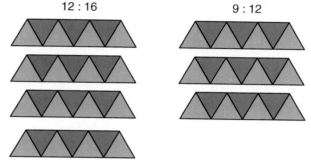

A **proportion** is a statement that two ratios are equal.
For example, 2 : 3 = 10 : 15
or, $\frac{2}{3} = \frac{10}{15}$

To **solve a proportion** means to determine the value of an unknown term in a proportion.
For example, to solve 10 : 15 = n : 3, determine the value of n that makes the ratios equal.

A **rate** is a ratio of two terms with different units.
A **unit rate** is a rate where the second term is 1 unit.
For example, 50 km : 1 h is written as 50 km/h.

A **percent** is a ratio with second term 100.
For example, 25 : 100 can be written:
in fraction form as $\frac{25}{100}$,
in decimal form as 0.25, and
as a percent, 25%

What Should I Be Able to Do?

4.1

1. Write 2 equivalent ratios for each ratio.
   a) 3:5
   b) 36:42
   c) 15:10
   d) 225:35

2. Elise made up a game with this spinner.

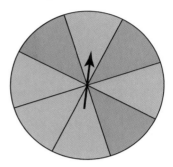

   If the pointer lands on red, the player loses a turn.
   Elise made a larger spinner. It had 20 sectors, 12 of which are red.
   a) Are the spinners equivalent?
   b) If your answer to part a is yes, justify your answer.
      If your answer is no, describe a spinner that would be equivalent to the first spinner.

3. Orange juice is mixed from concentrate and water.

   A
   - 4 parts concentrate
   - 5 parts water

   B
   - 3 parts concentrate
   - 4 parts water

   Which mix is stronger? Justify your answer.

4.2

4. Determine the value of each variable.
   a) $3:4 = 9:n$
   b) $a:35 = 4:20$
   c) $10:20 = 21:e$
   d) $20:m = 15:18$

5. A recipe for salad dressing calls for 4 parts oil to 1 part vinegar. Allie used 60 mL of vinegar. What volume of oil does she need?

6. A recipe that makes 5 dozen cookies calls for 4 eggs and 2 cups of flour.
   a) Wolfgang has only 3 eggs. How much flour should he use?
   b) How many cookies will he make?

4.3

7. Determine each unit rate.
   a) 240 km driven in 5 h
   b) 105 words typed in 3 min
   c) $2.80 for a 7-min call
   d) $4.74 for 3 kg of oranges
   e) 240 pages printed in 8 min

8. Which toothpaste is the better buy? Show your work.

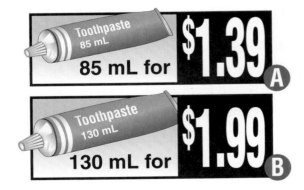

85 mL for $1.39 A
130 mL for $1.99 B

140 CHAPTER 4: Proportional Reasoning

4.4

9. The mass of grass seed needed to seed a yard depends on the area of the yard. Five kilograms of seed cover 100 m².
   a) How much seed is needed for this lawn?

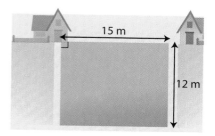

   b) Seed is sold in 8-kg bags. One bag of seed costs $21.20. How much will it cost to seed the lawn in part a?

10. Paige works after school picking apples. It took her 4 days to pick 9 rows of trees.
   a) At this rate, how long will it take her to pick 20 rows of trees? Explain.
   b) Paige picked for 10 days. How many rows did she pick?

4.5

11. Mr. O'Shea drove 600 km on 50 L of gas.
   a) How far could he drive on 30 L of gas?
   b) How much gas would he need to travel 420 km?
   c) What strategy did you use to solve each problem? Explain your strategy and why you chose it.

12. A photocopier can print 12 copies in 48 s. At this rate, how many copies can it print in 1 min?

4.6

13. Determine each percent.
   a) 20% of $56.99
   b) 45% of $118.56
   c) 30% of $89.99
   d) 25% of $37.88

14. A winter jacket is regularly priced at $79.99. It is on sale for 35% off.
   a) What is the sale price?
   b) What does the customer pay, including taxes?

15. The Canadian Radio and Television Commission requires 60% of the programming on CBC to be Canadian content. The CBC broadcasts from 6 a.m. one day to 2 a.m. the next each day. How many hours of programming each week are Canadian content?

16. Sheila put $350 in a savings account for 10 months. The annual interest rate was 3%.
   a) How much simple interest did the money earn?
   b) How much money was in the account after 10 months?

17. James borrowed $1500 for 8 months. The annual interest rate was 9%.
   a) How much simple interest did James pay?
   b) What did the loan cost James?

Practice Test

Multiple Choice: Choose the correct answer for questions 1 and 2.

1. The ratio of length to width of a rectangle is 5:3.
   Which dimensions could be those of the rectangle?
   A. 9 cm by 12 cm
   B. 16 km by 10 km
   C. 10 m by 6 m
   D. 9 mm by 16 mm

2. A 4-L can of paint covers an area of 32 m².
   What area will a 10-L can of paint cover?
   A. 40 m²
   B. 60 m²
   C. 80 m²
   D. 320 m²

Show your work for questions 3 to 6.

3. **Knowledge and Understanding**
   Determine the value of each variable.
   a) $2:10 = 5:a$
   b) $30:42 = b:7$
   c) $18:30 = 12:c$
   d) $\frac{n}{4} = \frac{21}{10}$

4. **Communication** Sun Li bought 12 oranges for $6.48.
   How many oranges could she buy with $10.00?
   How many different ways could you find out?
   Explain each way.

5. **Application** Jenna put $2500 in a savings account.
   The annual interest rate was 2%.
   a) How much simple interest did the money earn in 3 months?
   b) How much money was in the account after 3 months?

6. **Thinking** The sun is shining.
   You have a measuring tape and a friend to help you. How could you determine the height of a flagpole?
   Explain your strategy.

142 CHAPTER 4: Proportional Reasoning

Chapters 1–4 Cumulative Review

CHAPTER 1

1. Determine the perimeter and area of each figure.
   a)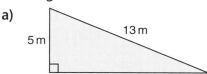
   b)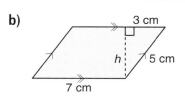

2. Determine the area and perimeter of this figure.
   The curve is a semicircle.
   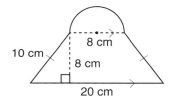

3. Determine the volume of each object.
   a)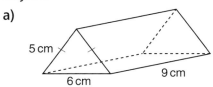
   b)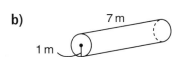
   c)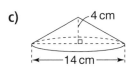
   d)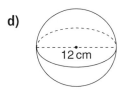

4. A pyramid has height 12 cm and base area 28 cm².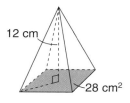
   a) What is the volume of the pyramid?
   b) What is the height of the related prism?
   c) What would the prism's height have to be for it to have the same volume as the pyramid? Check your answer.

CHAPTER 2

5. A rectangle has area 20 cm². Give 4 possible lengths and widths for the rectangle.

6. Kate has 18 m of edging for a rectangular garden.
   a) What are the dimensions of the largest garden Kate can enclose?
   b) What is the area of this garden?
   c) Suppose one side of the garden is along an existing fence and does not need edging.
      i) What are the dimensions of the largest garden that Kate can enclose?
      ii) What is its area?

7. a) For each area, determine the dimensions of a rectangle with the minimum perimeter.
      i) 25 cm²
      ii) 32 cm²
      iii) 78 m²
   b) Calculate the perimeter of each rectangle in part a. How do you know each perimeter is a minimum?

Cumulative Review **143**

CHAPTER 3

8. An isosceles triangle has one angle measuring 34°. What are the measures of the other 2 angles? Give two possible answers.

9. Determine the angle measure indicated by each letter. Justify your answers.

a)
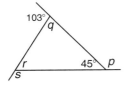

b)
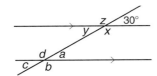

c)
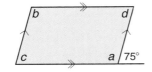

10. a) Use this diagram to determine the sum of the interior angles in a hexagon.

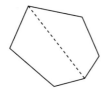

b) Show a different way that you could divide this hexagon to determine the sum of the interior angles. Determine the sum. How does it compare to your answer to part a?

c) A regular hexagon has 6 equal sides and 6 equal angles. Determine each measure.
  i) an interior angle
  ii) an exterior angle

CHAPTER 4

11. Write 2 equivalent ratios for each ratio.
a) 2 : 7   b) 45 : 165
c) 21 : 9   d) 100 : 24

12. A recipe that makes 6 dozen meringue drops calls for 4 egg whites and 1 cup of sugar.
a) Saad has only 3 eggs. How much sugar should he use?
b) How many meringue drops will Saad make?

13. Hannah's favourite cereal comes in 2 sizes. Which is the better buy? Show your work.
• 775 g for $5.49
• 400 g for $2.97

14. Running an old gas-powered lawn mower for 1 h can create as much pollution as driving a new car 550 km. It takes Isaac 45 min to mow his front and back lawns. How far would a person have to drive a new car to generate the same pollution?

15. Determine each percent.
a) 30% of $129.99
b) 15% of $27.54
c) 40% of $215.49

16. Hong put $550 in a savings account for 9 months. The annual interest rate was 4%.
a) How much simple interest did the money earn?
b) How much money was in the account after 9 months?

5 Graphing Relations

What You'll Learn
The relationship between two quantities can be illustrated with a graph that could be a straight line, a curve, or neither of these.

And Why
Relationships, such as the changes in temperature during a day or the time to run a certain distance, can be described with graphs. These graphs can be used to make predictions and to solve problems.

Key Words
- scatter plot
- trend
- line of best fit
- curve of best fit
- linear relation
- non-linear relation

Project Link

- Help Find the Stolen Mascot!

Reading and Writing in Math

Writing Solutions

When your teacher asks you to "show your work," you must show all your thinking and write a complete solution to the question.

One question in a text was:

"A cone has height 8 m and base radius 10 m.
 Determine the volume of the cone.
 Show your work."

Here are 2 students' solutions.

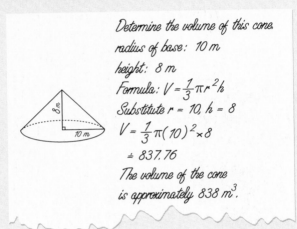

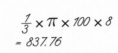

➤ The solution on the left is more detailed.
   List some things that are included in the detailed solution.
➤ Suppose you did not know what the question was.
   Could you follow the thinking of the student on the right?
   Explain.
➤ Which solution looks more like your solutions?

Here are some tips to write a solution:

- Write the question.
- Show all your steps so someone else can follow your thinking.
- Include graphs, tables, or diagrams if they help explain your thinking.
- Is your solution reasonable? If it does not make sense, check all your calculations.
- Use math symbols, such as $=$ and $+$, correctly.
- Include symbols, such as m and m^2, correctly.
- Write a sentence that answers the question.
- Include part of the question in the answer.

5.2 Line of Best Fit

To assess a person's nutritional needs, a doctor should know the person's height. But it can be difficult to measure the height of a person with a disability. Researchers look for other measures that can be used to estimate a person's height, such as knee height, arm length, and arm span.

Investigate   Trends in Measurement

Do you think a taller person has longer arms than a shorter person? Answer the question, then conduct the following experiment to check your prediction.

Work in a group of 3.
You will need a measuring tape.
Work together to measure each person's height and arm length.
Record these measurements on the board.
Measure each person's knee height.
Record your knee height and your height in your notebook for use in *Practice*, question 5, page 155.

Copy the data from the board into a table like this.
Draw a scatter plot.
Does there appear to be a relationship between height and arm length?
How do you know?
Compare your results with your prediction.
If they are different, explain why.

Height (cm)	Arm length (cm)

Keep your scatter plot for use in *Practice*, question 2, page 153.

Reflect

Use your scatter plot to answer these questions.
Justify your answers.

➤ A person is 160 cm tall. What might her arm length be?

➤ A person has arm length of 75 cm. How tall might he be?

Compare your answers with those of your classmates.
If the answers are different, explain why.

Connect the Ideas

The owner of an ice-cream stand wants to predict the number of ice-cream cones she will sell each day.
From her experience, she thinks the number may be related to the daily maximum temperature.
For 2 weeks, the owner records the maximum temperature and the number of cones sold each day.
She then draws a scatter plot.

Draw the scatter plot

Recall that this mark, ⩚, on an axis indicates that numbers are missing from the scale.

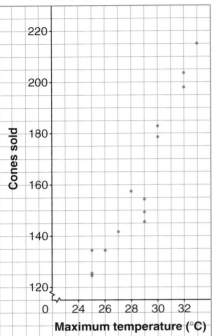

Ice-cream Cone Sales

The scatter plot shows a trend in the data. The points go up to the right. That is, as the daily maximum temperature increases, more cones are sold.

Draw the line of best fit

To help predict the number of cones that might be sold, we draw a **line of best fit**.

There should be about as many points above the line as below.

To do this, place a ruler on the graph so that it follows the path of the points. Draw a straight line along the edge of the ruler.

Use the line to estimate

One day, the predicted maximum temperature is 31°C. To estimate the number of cones that might be sold, use the line of best fit. Begin at 31 on the *Maximum temperature* axis. Move up to the line of best fit, then over to the *Cones sold* axis. When the temperature is 31°C, about 186 cones might be sold.

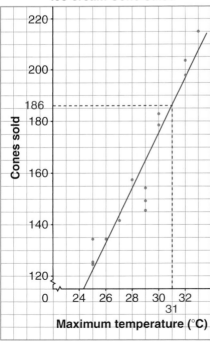

Ice-cream Cone Sales

152 CHAPTER 5: Graphing Relations

Practice

1. Adrianna drew a scatter plot and three different lines.
   Which line would you use as a line of best fit? Justify your choice.

   a) Wrist Circumference and Height

   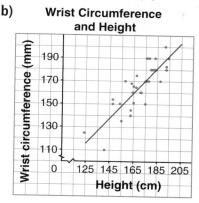

   b) Wrist Circumference and Height

   c) Wrist Circumference and Height

   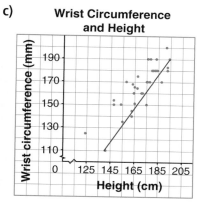

2. Use the scatter plot you drew in *Investigate*, page 151.
   a) Draw a line of best fit.
   b) Use your line to answer the two questions in *Reflect*, page 151.
   c) How do your answers in part b compare to your predictions in *Reflect*?
   d) Compare your predictions with those of two classmates.
      Does the line of best fit give closer predictions than the graph without the line? Explain.

3. Ask your teacher for a copy of the two scatter plots from Section 5.1, page 149.
   a) Only one of these scatter plots may be modelled with a line of best fit.
      Which is it? Explain why you should not draw a line of best fit for the other scatter plot.
   b) Draw a line of best fit for the plot you chose in part a.
      Use your line to make a temperature prediction for a latitude or elevation not included on the scatter plot.

You can extend a line of best fit to predict values beyond the data points.

Example

The table shows the world record times for women's 500-m speed skating from 1983 to 2001.

Year	1983	1986	1987	1987	1988	1994	1995	1997	1997	1997	2001	2001	2001
Time (s)	39.69	39.52	39.43	39.39	39.10	38.99	38.69	37.90	37.71	37.55	37.40	37.29	37.22

a) What trend do you see in the data?
b) Draw a scatter plot and a line of best fit.

c) What might the record have been in 1981?
d) When might the record time be less than 37 s?
e) What assumptions are you making in part d?

Solution a) The world record times are decreasing. That is, the skaters are getting faster.

b)
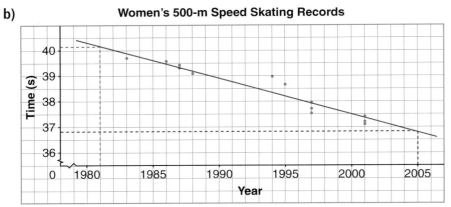

c) From the line of best fit, the world record in 1981 might have been about 40.2 s.
d) From the line of best fit, the world record might drop below 37 s by 2005.
e) We assume the skaters can keep improving their times.

4. A ball is dropped from different heights. The drop height and rebound height are recorded.

Drop height (m)	Rebound height (m)
1.0	0.7
2.0	1.3
3.0	2.3
4.0	3.0
5.0	3.8

   a) What trend do you see in the data? Does the rebound height of the ball appear to depend on the height from which the ball is dropped?
   b) Draw a scatter plot. Does the scatter plot support your answers to part a? Explain.
   c) Draw a line of best fit.
   d) Predict the rebound height when the ball is dropped from a height of 2.5 m and from a height of 8 m. How did you do this?
   e) Write one other question about these data. Answer your question. Show your work.

   Having Trouble? Read the Example above.

5. **Assessment Focus** In *Investigate*, page 151, you recorded your height and knee height.
   a) Do you think a person's knee height is related to her or his height? Explain.
   b) Use your measurements and those of 15 classmates.
   Draw a scatter plot.
   If possible, draw a line of best fit.
   Do the data support your answer to part a? Explain.
   c) Estimate the knee height of a person who is 180 cm tall.
   d) Estimate the height of a person with a knee height of 50 cm.

6. Is a person's shoe size related to her or his height?
   To check your answer, use the height data from *Investigate*, page 151, and collect shoe size data.
   a) Graph the data. Can you draw a line of best fit? Justify your answer.
   b) Do the data support your answer? Explain.

7. Choose a topic to investigate that involves two measures.
   Pose a question about the relationship between the measures.
   Conduct an experiment or research to find the data you need to answer the question.
   Present your solution using words, numbers, and graphs.
   If the data appear to lie along a line, include a line of best fit on the graph.

8. Will two people always draw the same line of best fit for a set of data? Explain.

9. **Take It Further** Conduct research to find the world record times for men's 500-m speed skating from 1983 to 2001.
   a) What trend do you see in the data?
   b) Draw a scatter plot and a line of best fit.
   c) Estimate what the record might have been in 1981.
   How did you do this? Do some research to check the estimate.
   d) Write a question that can be answered using the line of best fit.
   Prepare a solution for the question.
   Give the graph and question to a classmate to answer.
   Discuss and compare your solutions.

In Your Own Words

List 2 reasons you might draw a line of best fit after plotting data.
Provide an example for each reason.

Using *Fathom* to Draw a Line of Best Fit

As part of the international project called *Census at School*, Statistics Canada has collected data on Canadian high school students.

We will use *Fathom* to explore relationships in the data for 104 Ontario students. These students were selected from the *Census at School* database.

Here is a "collection inspector" window for the data set:

In *Fathom*, a collection is a data set.

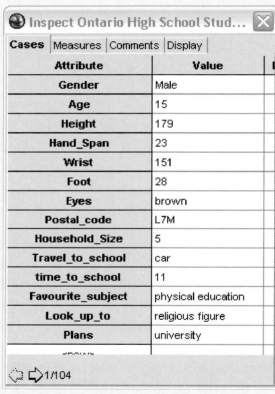

Source: Statistics Canada, *Education Matters: Insights on Education, Learning, and Training in Canada.* "Census at School". Catalogue no. 81-004-XIE, Vol. 2, No. 5, released February 28, 2006. Date extracted June 23, 2006.

We see that this person is a 15-year-old male who is 179 cm tall, has a hand span of 23 cm, a wrist circumference of 151 mm, a foot length of 28 cm, and brown eyes.

We can also see information about his home and his life. Notice that some attributes, such as height, are described by a number.

Other attributes, such as plans, are described in words.

We will create scatter plots to help us look for relationships between attributes described by numbers.

Use *Fathom* to open the file *Ontario High School Students 1*. This is a ready-made data set on 104 students.	
You will see a scatter plot showing hand span on the horizontal axis and foot length on the vertical axis. We see that, generally, the greater a person's hand span, the longer her or his foot.	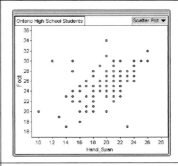
To model this relationship with a line: Click on the graph so it becomes active. The **Graph** menu appears on the toolbar at the top of the screen. Select the **Graph** menu. Then select **Movable Line**.	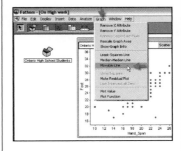
"Fit" the brown line to the data so it illustrates the relationship between foot length and hand span. You can "grab and drag" the ends of the brown line to make it steeper or less steep. You can "grab and drag" the middle of the brown line to move it up or down.	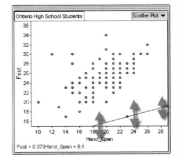
When the line fits the data the best, the data points should be equally distributed on both sides of your line of best fit. Notice that *Fathom* gives you the equation for your line of best fit. Print the graph.	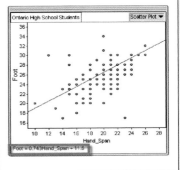

Both the hand span and foot length are measured in centimetres.

Double-click the collection. The first case in the collection is displayed beside the graph.	
To explore if there is a relationship between hand span and wrist circumference: Left click, drag, and drop the "Wrist" attribute from the collection inspector to the vertical axis of your scatter plot. The foot length data are replaced by the wrist circumference data.	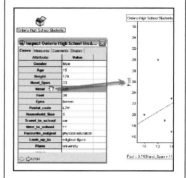
Because the vertical axis now begins at a greater number, the brown movable line is no longer visible. "Grab" the bottom of the vertical axis and drag it up, until the movable line appears. "Grab and move up" the line. You can "grab and push" the vertical axis back where it was.	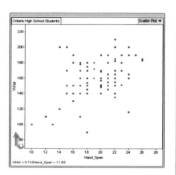
"Fit" the movable line to the data. Describe the relationship between hand span and wrist circumference. Do you think it is appropriate to draw a line of best fit for these data? Justify your answer. Print your graph.	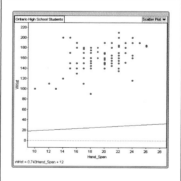
Insert a text box. Type in your name. Save your file. Try to find relationships between other attributes described by numbers.	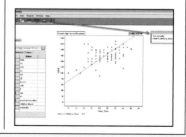

For each relationship you investigated, compare your line of best fit with those of other students. How are they similar?

5.3 Curve of Best Fit

Investigate — Relationship between Temperature and Time

Work in a group of 3.
You will need: a plastic cup, an insulated cup, crushed ice, water, a measuring cup, and a thermometer

Place equal volumes of ice and warm water in each cup.
Predict what will happen to the ice and water in each cup.
Predict how the temperature will change over time.
Give reasons for your predictions.

Measure the temperature of the contents of each cup at regular intervals for 30 min.
How often do you think you should measure the temperatures? Why?
Record the times and temperatures in a table.

Use a different colour to plot each set of data on the same grid.

How are the graphs the same?
How are they different?
Describe any trends in the graphs.
Do the results support your predictions? Explain.

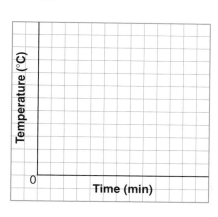

How long did it take the contents of each cup to reach a constant temperature?

Reflect

➤ What factors might affect the results of this experiment?

➤ How could you change this experiment to account for these factors?

Compare your results with those of your classmates.
If they are different, explain why.

Connect the Ideas

A weather forecaster measures and plots the temperature every 2 h on a summer day.

The temperatures decrease from midnight to 4:00 a.m., increase once the sun rises, reach a maximum in the afternoon, then decrease again during the late afternoon and evening.

The points do not lie on a straight line, but appear to be related. These data can be approximated by a curve.

We call it a **curve of best fit**.

To draw a curve of best fit, draw the smooth curve that passes through as many points as possible.

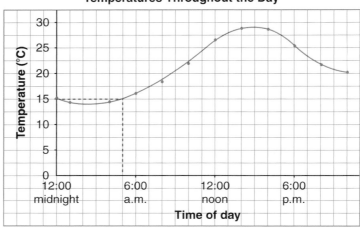

The greatest temperature occurred between 2:00 p.m. and 3:00 p.m.
We can use the curve to estimate the temperature at 5:00 a.m.
Begin at 5:00 a.m. on the *Time of day* axis.
Move up to the curve, then across to the *Temperature* axis.
At 5:00 a.m., the temperature was about 15°C.

Practice

1. Your teacher will give you a copy of each graph.
Describe any trends in the data. Draw a curve of best fit for each set of data.

a) Path of an Arrow

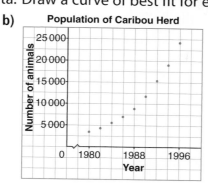

b) Population of Caribou Herd

c) Growth of a Sunflower

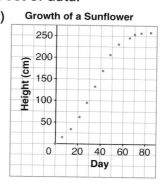

2. The table shows the number of hours of daylight in Waterloo, Ontario, for the first day of each month in 2005.

Month	Jan	Feb	Mar	April	May	June	July	Aug	Sep	Oct	Nov	Dec
Daylight hours	9.0	9.9	11.2	12.8	14.2	15.2	15.4	14.5	13.2	11.7	10.3	9.2

a) What trend do you see in the data? Explain the trend.
b) Graph the data. Draw a curve of best fit.
c) Estimate the number of hours of daylight on March 15.
d) The day with the most daylight is June 21.
Estimate the number of hours of daylight on June 21.
e) Estimate the number of hours of daylight on your birthday.
How did you do this?

You have used a curve of best fit to predict values that lie between data points.
You can also extend a curve of best fit to predict values beyond the data points.

Example

A soccer ball is kicked up into the air from the ground.
The height of the ball is measured at regular time intervals.
Here are the data.

Time (s)	0	0.2	0.4	0.6	0.8	1.0	1.2	1.4	1.6
Height (m)	0	2.2	4.0	5.4	6.5	7.1	7.3	7.2	6.7

a) What trend do you see in the data? Explain the trend.
b) Graph the data. Draw a curve of best fit.
c) When do you think the ball is at its greatest height?
Use the graph to check.

5.3 Curve of Best Fit

d) When is the ball 5 m high?

e) When does the ball hit the ground? How does the graph show this?

Solution
a) The height of the ball increases, then decreases. That is, the ball stops rising and begins to fall.

b)
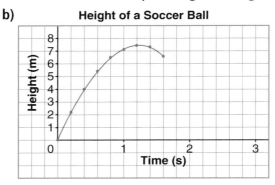

c) From the table and graph, the greatest height of the ball seems to occur at 1.2 s when the ball is 7.3 m high.

d) There are two times when the ball is 5 m high: once as it is rising and once as it is falling.
To determine the second time, extend the graph to the right. Draw a smooth curve that "mirrors" the curve from the starting point to its greatest height.

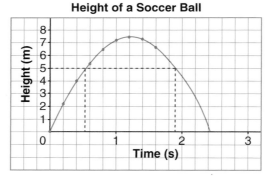

From the graph, the ball is 5 m high at approximately 0.5 s and 1.9 s.

e) The extended curve shows that the ball hits the ground after about 2.4 s, when the height is 0 m.
This is where the curve meets the *Time* axis.

3. An Internet host is a computer directly connected to the Internet. The number of Internet hosts around the world has grown quickly. The data in the table are for January of each given year.

Year	1992	1994	1996	1998	2000	2002	2004	2006
Internet hosts (millions)	0.7	2.2	9.5	29.7	72.4	147.3	233.1	395.0

a) What trend do you see in the data? Explain the trend.
b) Graph the data. Draw a curve of best fit.
c) Estimate the number of Internet hosts in 2001.
d) When might the number of Internet hosts reach 500 million? Justify your answer.

4. **Assessment Focus** The high divers at Paramount Canada's Wonderland perform competitive dives from a height of 21 m. A diver's height is measured every 0.2 s.

Time (s)	0	0.2	0.4	0.6	0.8	1.0	1.2
Height (m)	21	20.7	20.2	19.4	17.6	16	14

a) What trend do you see in the data? Explain the trend.
b) Graph the data. Draw a curve of best fit.
c) Estimate when the diver will be 10 m above the water.
d) When does the diver reach the pool? How do you know?

5. **Take It Further** The table shows the number of people enrolled in apprenticeship programs in Canada, rounded to the nearest thousand.

Year	1995	1996	1997	1998	1999	2000	2001	2002
Females (thousands)	11	12	13	14	16	17	20	22
Males (thousands)	153	154	159	163	173	184	198	213

a) Graph both sets of data on one grid.
b) For each data set, draw a line or curve of best fit. How did you decide which to draw?
c) Estimate the numbers of females and males enrolled in apprenticeship programs in 2003. How did you do this?

In Your Own Words

Suppose you graph data for which a curve of best fit can be drawn. How do you decide where to draw the curve? Include a graph in your explanation.

Using a Graphing Calculator to Draw a Curve of Best Fit

When data points appear to lie along a curve, we say the relation is **non-linear**.
There are many mathematical models to describe non-linear relations.
Two models are a quadratic curve of best fit and an exponential curve of best fit.

The TI-83 and TI-84 graphing calculators can model a relation with
- a line of best fit (**LinReg**)
- a quadratic curve of best fit (**QuadReg**)
- an exponential curve of best fit (**ExpReg**)

Three sample scatter plots are shown.
Each plot represents a different type of relation.

> It can be difficult to distinguish between quadratic and exponential relations. The quadratic curve increases more gradually.

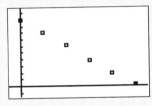   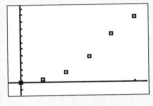   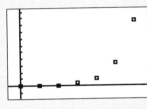

Linear   Quadratic   Exponential

A ball is dropped from a height of 3.5 m.
The height of the ball as it drops is measured using a CBR motion detector.
The data are shown in this table.

Time (s)	0.0	0.2	0.4	0.6	0.8
Height (m)	3.5	3.2	2.8	1.5	0.2

Use a TI-83 or TI-84.
Follow these steps to model this relation.

CHAPTER 5: Graphing Relations

Press [STAT] [1] to access the List Editor. If necessary, clear the lists L1 and L2. To do this, move the cursor to the column head and press [CLEAR] [ENTER]. Enter the times in L1, and the corresponding heights in L2. After you key in each number, press [ENTER]. Before graphing the data, make sure there are no equations in the [Y=] list. To do this, press [Y=]. Move the cursor to any equation, then press [CLEAR].

Press [2nd] [Y=] [1] to access the Stat Plot Editor for Plot 1. Set the plot options as shown here. Then press [ZOOM] [9]. This instructs the calculator to set the window so all the data can be displayed, and graphs the data.

The graph should look like the one shown here. The points do not lie on a line. We will model the data with an exponential curve and a quadratic curve, and decide which is the better fit.

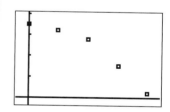

To draw an exponential curve of best fit, press [STAT] [▶] [0] [ENTER]. The calculator displays the equation of an exponential curve that models the data.

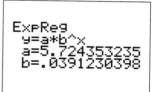

To show the curve, press [Y=] [CLEAR]. Press [Y=]. Then press [VARS] [5] [▶] [▶] [1] [GRAPH].

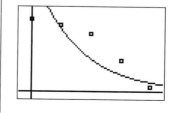

Technology: Using a Graphing Calculator to Draw a Curve of Best Fit

To draw a quadratic curve of best fit, press STAT ▷ 5 ENTER. The calculator displays the equation of a quadratic curve that models the data.	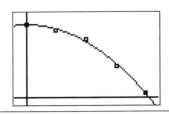
To show the curve, press Y= CLEAR. Then press VARS 5 ▷ ▷ 1 GRAPH.	

Which model better represents the data? Justify your choice.

Complete these steps for each set of data below.

Use pages 165 and 166 as a guide.

➤ Use a TI-83 or TI-84 to graph the data.

➤ Draw an exponential curve of best fit and a quadratic curve of best fit. Sketch each curve.

➤ Decide which curve better models the data. Justify your answer.

1. In 2002, movie attendance in Canada hit a 44-year high. The table shows the attendance in several other years for comparison.

Year	1991	1998	1999	2000	2002
Movie attendance in Canada (millions)	69.2	109.7	117.4	117.6	124.2

2. In ideal conditions, *E. coli* bacteria cells divide every 20 min. The table shows how the number of cells in a sample would grow over time.

Time (min)	0	20	40	60	80	100	120
Number of cells	1	2	4	8	16	32	64

Hidden Sum

Play in a group of 3.
You will need a calculator.
As a group, choose a target number greater than 200.

The first player enters a number into the calculator,
presses the Memory Plus key [M+],
and passes the calculator to the second player.

The next player enters a number and presses the [M+] key.
The number is added to the number already stored in the memory.

Take turns entering numbers and pressing the [M+] key.

When you think the sum is equal to or greater than
the target number, call out "Over!"
Use the Memory Recall key [MR] to check.
If you are correct, you win the game.
If not, you are out of the game.
The other players continue taking turns.

Each number entered must be less than 50.

Mid-Chapter Review

5.1 1. Jordan and Esau compared the statistics of the junior boys' basketball team. They drew a scatter plot.

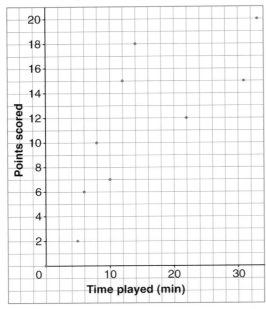

a) What does each point represent?
b) How many points were scored by the boy who played 10 min?
c) How many boys scored 15 or fewer points? More than 15 points?
d) Suppose a boy played few minutes but scored many points. Where would his point on the graph be?
e) Describe any trends in the data. Justify your answer.

5.2 2. Saskia and Maria are working on a science fair project. They investigate whether a dog's mass is related to its height.

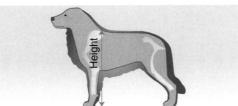

Here are some data Saskia and Maria collected.

Height (cm)	69	59	54	46
Mass (kg)	35	32	23	17

Height (cm)	38	31	28	27
Mass (kg)	14	6	8	5

a) Describe any trends in the data.
b) Graph the data.
   Draw a line of best fit.
c) Does there appear to be a relationship between a dog's height and mass?
   Justify your answer.
d) Estimate the height of a dog with mass 16 kg.
e) Estimate the mass of a dog that is 50 cm high.
f) What assumptions did you make in parts d and e?

5.3 3. This table shows the world population every 10 years from 1900 to 2000. The populations are rounded.

Year	1900	1910	1920	1930	1940	1950	1960	1970	1980	1990	2000
Population (billions)	1.6	1.8	1.9	2.1	2.3	2.6	3.0	3.7	4.5	5.3	6.1

a) Describe any trends in the data.
b) Graph the data.
   Draw a curve of best fit.
c) Estimate the population in 1925 and in 1975.
d) Estimate the population in 1890.
e) Predict the population in 2010.
f) What assumptions did you make in parts c, d, and e?

Practice

1. Two graphs are shown. Is each relationship linear or non-linear? Explain how you know.

a)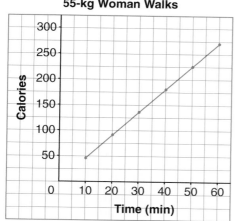

b)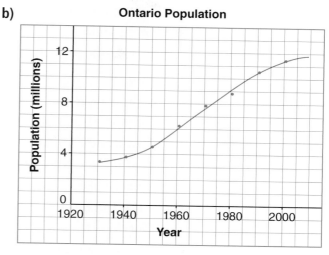

2. Use the graphs in question 1.
   a) i) About how many calories does a 55-kg woman burn when she walks for 30 min? For 45 min?
   ii) What was the approximate population of Ontario in 1945? In 1995?
   b) Write a question of your own that can be answered using one of the graphs.
   Exchange questions with a classmate. Answer your classmate's question.

3. In Chapter 2, you investigated the dimensions and perimeter of a rectangle when its area was given. Here are some data for a rectangle with area 36 cm².

Width (cm)	Length (cm)
1	36
2	18
3	12
4	9
6	6

a) Is the relationship between length and width non-linear? Justify your answer.
b) Graph the data.
   Does the graph illustrate your answer to part a? Explain.
c) Use the graph.
   i) Determine the length when the width is 5 cm.
   ii) Determine the width when the length is 8 cm.
d) Write a rule for the relationship.

5.5 Graphing Non-Linear Relations **177**

We can create a table of values from a description of a relationship.

Example

A typical North American adult consumes about 200 mg of caffeine a day. Caffeine has a half-life of about 6 h. This means that about 6 h after consumption, half the caffeine remains in a person's body.

> Caffeine occurs naturally in coffee beans, cocoa beans, kola nuts, and tea leaves. It is also found in products, such as coffee, tea, some energy drinks and soft drinks.

a) Copy and complete the table below to show how much caffeine is left in a person's body over time.

b) What trends do you see in the data? What do you think the graph will look like?

c) Graph the data. Describe the graph. How does the graph compare to your prediction in part b?

d) About how much caffeine will remain after 9 h? After 36 h? What assumption did you make?

Time (h)	Mass of caffeine (mg)
0	200
6	
12	
18	
24	

Solution

a) Every 6 h, the mass of caffeine is halved. The times in the table increase by 6 h each time. So, divide the mass by 2 to get the new mass each time.

b) The mass of caffeine is decreasing. The amount by which the mass changes is also decreasing. So, the graph will be a curve.

Time (h)	Mass of caffeine (mg)
0	200
6	100
12	50
18	25
24	12.5

c) Draw the curve of best fit through the points. The graph is a curve that goes down to the right.

d) From the graph, the mass of caffeine remaining after 9 h is about 70 mg. By extending the graph, the mass of caffeine after 36 h is about 5 mg.

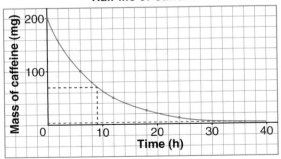

> We can also determine this mass by extending the table.
> After 30 h, the mass is 6.25 g.
> After 36 h, the mass is 3.125 g.
> We get an exact answer using the table.
> We assume that the person is not consuming any more caffeine during this time.

4. Use 1-cm grid paper.
   a) Draw squares with side lengths from 1 cm to 6 cm.
   b) Calculate the area of each square.
      Copy and complete this table.
   c) Graph the data.
      Describe any trends in the graph.
   d) Estimate the area of a square with side length 4.5 cm.
   e) Estimate the area of a square with side length 7.5 cm.
   f) How could you check your answers to parts d and e?

Side length (cm)	Area (cm²)
0	
1	

5. **Assessment Focus** The first 3 solids in a pattern are shown.

    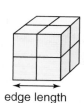 

   edge length

   a) Describe the pattern in the cubes.
   b) Sketch the next 2 solids in the pattern.
      Copy and complete this table.
   c) Describe any trends in the data.
   d) Suppose you know the edge length of a solid in this pattern.
      i) How can you determine the number of cubes needed to build it?
      ii) How is this number related to the volume of the solid?
   e) Add rows to your table for the next 2 solids in the pattern.
   f) Graph the data.
   g) How is the volume of a solid related to its edge length?
      Write a rule.
   h) What is the edge length of the solid with 512 cubes?
   i) How many cubes would be needed for the 10th solid?
   j) How many different ways could you answer parts h and i?

Edge length	Number of cubes
1	
2	

5.5 Graphing Non-Linear Relations

6. A new car is purchased for $23 000.
   Its value depreciates by 15% each year.
   The estimated value of the car over time is shown.

Year	Value of car ($)
0	23 000
1	19 550
2	16 618
3	14 125
4	12 006
5	10 205

   a) Predict the shape of the graph. Justify your prediction.
   b) Graph the data.
   c) Describe the graph.
      How does the graph compare to your prediction in part a?
   d) Estimate when the value of the car will be about $7500.
      How did you do this?

7. In *Investigate*, page 175, you explored the effect of changing the length of a pendulum.
   a) Do you think the time it takes for a pendulum to complete 6 swings is related to the mass of the object used?
      If your answer is yes, describe how they might be related.
   b) Design an experiment you could conduct to test your prediction.
      Carry out your experiment. Was your prediction correct?
      If you cannot complete the experiment, research to check your prediction.

8. Choose 2 measurements that you think are related.
   a) Pose a question about the measurements.
   b) Collect data to answer the question. Record the data in a table.
   c) Graph the data. Describe any trends in the data.
   d) Answer the question you posed.

   Use *Fathom* or a graphing calculator, if available.

9. Your teacher will give you a table that shows the mass of caffeine in some foods and drinks.
   a) Estimate your total caffeine intake for a typical day.
   b) Create a table and graph like those in the *Guided Example*, pages 178-179.
      Show the caffeine that remains in your body over time.
      Use your answer from part a as the initial value.
   c) Repeat parts a and b using data for a friend or family member.

10. **Take It Further** Recall that the formula for the volume of a cone is $V = \frac{1}{3}\pi r^2 h$.
    Suppose the height of the cone stays the same, but its radius changes.
    Predict if the relationship between the radius and volume is linear or non-linear.
    Investigate your prediction. Write what you find out.

In Your Own Words

How can you tell from data that they represent a non-linear relation?
How can you tell from a graph? Include examples in your explanation.

5.6 Interpreting Graphs

Graphs are often used to display information in the media.
These graphs are sometimes misleading or can be misinterpreted.
Knowing how to interpret graphs is an important media literacy skill.

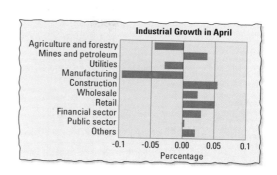

Investigate Representing Motion on a Graphing Calculator

Work in a group of 3.
Your teacher will give you instructions for using the CBR.
You will need: a TI-83 Plus or TI-84 graphing calculator, a CBR motion detector, and a calculator link cable.

You will use the CBR motion detector to investigate changes in your distance from a wall as you move toward and away from the wall. Connect your graphing calculator to the CBR using a link cable.

➤ Stand about 3 m away from a wall. Point the CBR at the wall.
When the CBR starts clicking, walk toward the wall,
stop for a few seconds, then walk away from the wall.
Have another group member record a description of your walk.
Sketch the graph displayed on the calculator.
What does the vertical axis represent?
What does the horizontal axis represent?

➤ Repeat the activity.
This time, stand close to the wall, walk away from the wall slowly, then quickly. Describe how the graph changes.

Reflect

Exchange both sketches of your graphs with those of another group.

➤ Describe the motion represented by each graph.

➤ Discuss the descriptions the group has written.
Were the descriptions accurate? Explain.

➤ What do the shapes of the graphs tell you about the motion?

Connect the Ideas

The graph shows the height of water in a bathtub over time. Key points where the graph changes are labelled.

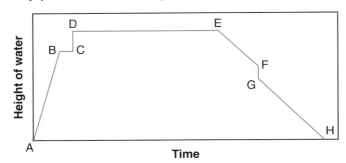

If we think about the possible reasons for the changes at the key points, we can describe what the graph represents.

At point A, the tub is empty.
A person puts in the plug and turns on the water.
From A to B, the tub fills with water.
At point B, the person turns off the water.
At point C, the person gets into the tub, causing the water level to rise suddenly.
The person sits in the tub from point D to point E.
At point E, he pulls out the plug and the water begins to drain.
At point F, the person gets out of the tub, causing the water level to drop suddenly.
From point G to point H, the water continues to drain from the tub.
At point H, the tub is empty.

Practice

1. Which graph best represents each situation? Explain your choices.
   a) The height of a baseball thrown up into the air measured over several seconds
   b) The height of a child measured over several years
   c) The height of a lit candle measured over several hours

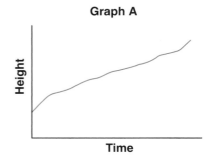
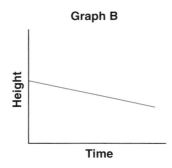
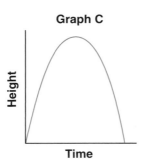

2. Julie and Osa were experimenting with a CBR and made these graphs. Describe their motions.

a)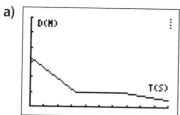

b)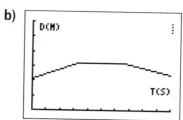

c)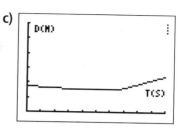

When a graph includes numerical data, we can describe what it shows in more detail.

Example

The graph shows Jorge's distance from home as he walks to school. Describe his walk.

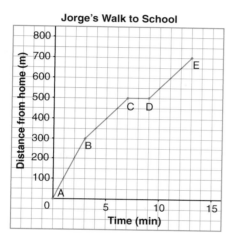

Solution From A to B, Jorge walks away from home.
The graph goes up to the right.
Jorge is walking for 3 min. He walks 300 m.

From B to C, Jorge is still walking away from home.
The segment from B to C is less steep than that from A to B.
So, Jorge's average speed has decreased and he is walking slower.
He walks for 4 min and travels 200 m.

From C to D, the segment is horizontal.
This means that time is passing, but Jorge's distance
from home stays the same. Jorge is standing still for 2 min.

From D to E, Jorge continues his walk to school at
about the same average speed as from B to C.
He walks for 4 min and travels 200 m.

It takes Jorge 13 min to walk 700 m to school.

3. The graph shows Olivia's distance from home as she walks to the store and back. Describe her walk.

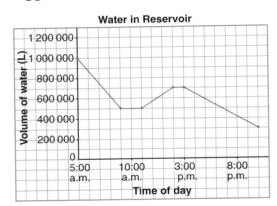

4. **Assessment Focus** The graph shows how the volume of water in a town reservoir changes during a typical day. Describe how the volume of water changes during the day. Suggest reasons for the changes.

5. **Take It Further** Aaya measured the temperature of a container of ice water as it was heated. She drew this graph. Describe how the temperature changes over time. Suggest reasons for the changes.

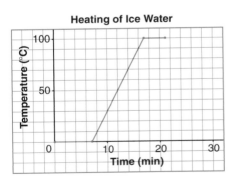

In Your Own Words

How do the graph title and the axes labels help you interpret data on a graph? Include an example in your explanation.

Chapter Review

What Do I Need to Know?

Line of Best Fit

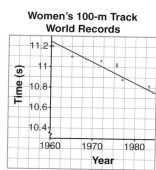

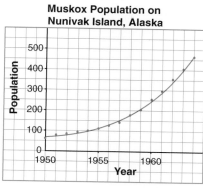

Curve of Best Fit

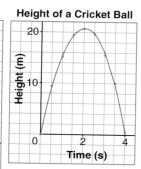

➤ A linear relation has a graph that is a straight line.
For example, the perimeter of a rectangle with width 3 cm:

Length (cm)	Perimeter (cm)
4	14
5	16
6	18
7	20

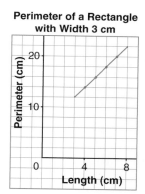

Use a Frayer model to help you understand some of the key words on this page.

➤ A relation that does not have a straight line graph is non-linear.
For example, the half-life of caffeine in the human body:

Time (h)	Mass of caffeine (mg)
0	300
6	150
12	75
18	37.5
24	18.75
30	9.375

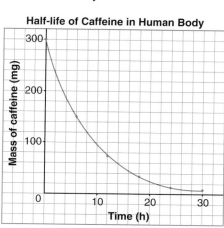

What Should I Be Able to Do?

5.1   **1. a)** What does this scatter plot show?

Prices of a Certain Model of Used Car

(scatter plot: Price ($) vs Age (years), with prices up to 20 000 and ages up to 15)

    b) How old is the car that costs $16 500?

    c) Data are included for two 6-year-old cars.
What are their prices?
Why do you think the prices are different?

    d) Describe any trends in the data. Explain your thinking.

5.2
5.3   **2.** Joseph and Malik measured the growth of a seedling.

Time (days)	1	3	5	7	9
Height (mm)	4	12	22	36	45

    a) Graph the data. Describe any trends.

    b) Draw a line or curve of best fit. Explain how you decided which to draw.

    c) i) How tall do you think the seedling was after 4 days?

      ii) How tall might it be after 10 days? How do you know?

    d) About how many days might it take for the seedling to reach a height of 60 mm?

    e) Can you predict how tall the seedling might be after 1 year? Explain.

3. The Canadian Recording Industry Association keeps track of sales of recorded music in Canada. Data for 1996 to 2003 are shown in the graph.

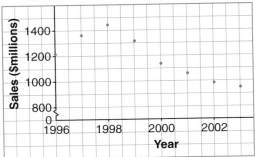

Retail Sales of Recordings in Canada

    a) Describe any trends in the data.

    b) Your teacher will give you a copy of the graph. Draw a curve of best fit.

    c) From the graph, about how much did Canadians spend on recorded music in 1998? In 2003?

    d) Predict about how much Canadians spent on recorded music in 2004. What assumptions are you making?

186   CHAPTER 5: Graphing Relations

5.2
5.3

4. A mug of hot chocolate was left on a counter to cool. Its temperature was measured every 2 min.

The results are shown in the table.

Time (min)	0	2	4	6	8	10	12	14	16
Temp (°C)	90	86	81	78	74	72	70	68	66

a) Graph the data.
   Describe any trends.
b) Draw a line or curve of best fit. Explain how you decided which to draw.
c) A separate mug of hot chocolate was poured at the same time. After 1 min some cold milk was added.
   How will the graph for this situation differ from the graph you drew in part a?
   Sketch the new graph.

5.4

5. The first 3 frames in a pattern are shown.

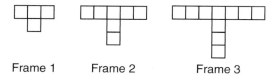

Frame 1   Frame 2   Frame 3

a) Draw the next 3 frames in the pattern.
b) For each of the 6 frames, record the frame number and the number of squares in a table.
c) Describe the trend.
   Do you think the graph will be linear or non-linear?

d) Graph the data. Was your prediction in part c correct?
e) Write a rule for the number of squares in any frame.
f) How many squares will there be in the 8th frame?
   How do you know?
g) Which frame will have 28 squares?
   How do you know?

5.5

6. Matthew and Tamara are conducting an experiment.
They vary the resistance of a resistor and measure the current through it.
Their results are shown in the table.

Resistance (ohms)	10	20	30	40	50	60
Current (A)	12	6	4	3	2.4	2

a) Describe the trend.
b) Graph the data.
c) What will the current be when the resistance is 15 ohms?
d) What will the resistance be when the current is 10 A?

5.6

7. The graph shows Hasieba's distance from home during a walk to the park.
Describe her walk.

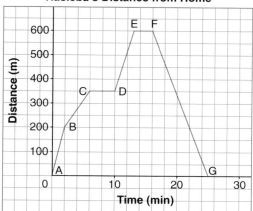

Chapter Review **187**

Practice Test

Multiple Choice: Use the graph. Choose the correct answer for questions 1 and 2.

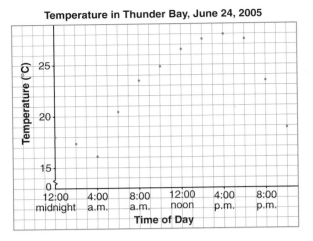

1. When was the temperature greatest?
   A. 5 p.m.
   B. 2 p.m.
   C. 4 p.m.
   D. 11 a.m.

2. Which is the best estimate for the temperature at 7 a.m?
   A. 18°C
   B. 20°C
   C. 22°C
   D. 25°C

Show your work for questions 3 to 6.

3. **Application** Water exerts pressure on a scuba diver. The pressure is measured in units called kilopascals (kPa). The table shows the approximate pressure at different depths of sea water.

Depth (m)	Pressure (kPa)
5	150
10	200
15	250
20	300
25	350

   a) Graph the data. Draw a curve or line of best fit as appropriate. Explain your choice.
   b) At what depth is the pressure 225 kPa? 400 kPa?
   c) What is the pressure at the surface of the water?
   d) Describe the relationship between depth and pressure.

4. **Knowledge and Understanding** A ball is dropped. The height it reaches after each bounce is measured. The data are shown in the table.

Bounces	1	2	3	4	5	6
Height (m)	2.1	1.5	1	0.7	0.4	0.3

   Which tools could you use to help you?

   a) Describe the trend in the data.
   b) Graph the data. Draw a curve or line of best fit as appropriate. Explain your choice.
   c) From what height do you think the ball was dropped? Explain your answer.
   d) What else could you find out from the graph?

5. **Thinking** William and Rhiannon run 200 m. Describe each person's run. Who finishes first? Justify your answer.

   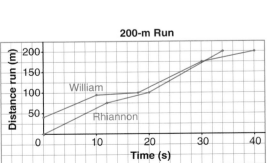

6. **Communication** When you look at a graph, how can you tell if it represents a linear relation, a non-linear relation, or neither? Include diagrams in your answer.

188 CHAPTER 5: Graphing Relations

6 Linear Relations

What You'll Learn
How linear relations connect to earlier work in this text on proportional reasoning, and geometry and measurement concepts

And Why
Real-life situations can be represented as linear relations in different ways. This helps to show how math relates to life outside school.

Key Words
- first differences
- rise
- run
- rate of change
- direct variation
- partial variation
- vertical intercept

Project Link

- Heart Rates, Breathing Rates, and Exercise

Making a Mind Map

Organizing information can help you learn and remember the important ideas.
A *mind map* is one way to do this.

Here is a student's mind map on fractions.

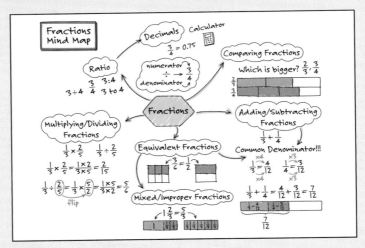

To make your own mind map:

➢ In the centre of a piece of paper, write the chapter title:
Linear Relations

➢ As you work through this chapter, list all the important ideas.
Make sure to include:
  • section titles, such as *Direct variation* and *Partial variation*
  • new words
  • key ideas

➢ At the end of each section, compare your list with a classmate's list. Arrange your ideas on your map.
Look for connections among the ideas.
Organize your map to highlight these relationships.

➢ At the end of the chapter, make a final copy of your map.
Use colour and diagrams to highlight and explain important concepts. Use arrows to link ideas.

➢ Compare your map with those of your classmates.
Talk with a classmate.
  • How could you have organized the information differently?
  • What was effective about each map?
  • How could you improve your mind map?

6.1 Recognizing Linear Relations

Investigate — Using a Table of Values to Explore a Relationship

Work in a group of 4.

➤ You will need linking cubes.
Place a cube on the desk.
Five faces of the cube are visible.
Join 2 cubes. Place them on the desk.
Eight faces of the cubes are visible.
Record the data in a table.
Continue the table to show data for
up to 10 cubes in a line.

Number of cubes	Number of faces visible

What patterns do you see in the table?
Graph the data. How are the patterns shown in the graph?
Is the relationship linear or non-linear? How do you know?

➤ Consider this pattern of cubes.

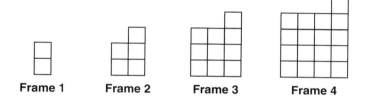

Frame 1   Frame 2   Frame 3   Frame 4

Use linking cubes to build these frames.
How many cubes are in each frame?
Record your data in a table.
Continue the table to show data for
6 frames.

Frame number	Number of cubes

What patterns do you see in the table?
Graph the data. How are the patterns shown in the graph?
Is the relationship linear or non-linear? How do you know?

> **Reflect**
> Suppose you have a table of values.
> How can you tell if the relationship it represents is linear or non-linear without drawing a graph?

Connect the Ideas

Look at a pattern

Use square tiles, grid paper, or square dot paper to help you picture the pattern.

Here is a pattern of squares. The pattern continues.

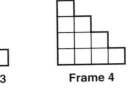

Frame 1   Frame 2   Frame 3   Frame 4

Record the data

Here are tables for the perimeter and area of each frame.

Frame number	Perimeter (units)
1	4
2	8
3	12
4	16
5	20
6	24

Frame number	Area (square units)
1	1
2	3
3	6
4	10
5	15
6	21

Notice that the frame numbers increase by 1 each time.

Graph the data

Here are the graphs of the data.

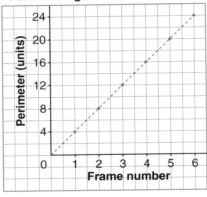
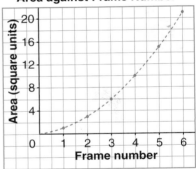

Analyse the data

The graph of *Perimeter* against *Frame number* shows a linear relation. The points lie on a straight line. We join the points with a broken line to show the trend.

The graph of *Area* against *Frame number* shows a non-linear relation. The points lie along a curve. We join the points with a broken curve to show the trend.

The broken line and broken curve indicate that only plotted points are data points.

First differences for a linear relation

Add a third column to the Perimeter table.
Record the changes in perimeter.

Frame number	Perimeter (units)	Change in perimeter (units)
1	4	
2	8	8 − 4 = 4
3	12	12 − 8 = 4
4	16	16 − 12 = 4
5	20	20 − 16 = 4
6	24	24 − 20 = 4

The numbers in the third column are called **first differences**.
The first differences for a linear relation are equal.
For every increase of 1 in the frame number,
the perimeter increases by 4 units.

All linear relations have first differences that are constant.
We can use this to identify a linear relation from its table of values.

First differences for a non-linear relation

Add a third column to the Area table.
Record the changes in area.

Frame number	Area (square units)	Change in area (square units)
1	1	
2	3	3 − 1 = 2
3	6	6 − 3 = 3
4	10	10 − 6 = 4
5	15	15 − 10 = 5
6	21	21 − 15 = 6

For a non-linear relation, the first differences are not equal.
For every increase of 1 in the frame number,
the area increases by a different amount each time.

6.1 Recognizing Linear Relations

Practice

1. For each table below, determine the first differences.
   State whether the data represent a linear or non-linear relation.
   Explain how you know.

   a)
Time rented (h)	Cost ($)
0	0
1	6
2	12
3	18
4	24

   b)
Age (months)	Mass (kg)
0	3.0
1	4.0
2	4.8
3	5.5
4	6.2

   c)
Time (s)	Distance (m)
0	15
1	20
2	25
3	30
4	35

2. Suggest a possible situation that each table in question 1 could represent.

3. This pattern was constructed using square tiles:

   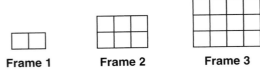

   Frame 1   Frame 2   Frame 3

   a) Copy and complete this table.
      Record data for the first 6 frames.
   b) Determine the first differences.
      Is the relationship between frame number and area linear?
      Explain.
   c) Graph *Area* against *Frame number*.
      Does the graph support your answer to part b? Explain.

Frame number	Area (square units)

4. This pattern was constructed using square tiles:

   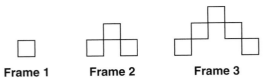

   Frame 1   Frame 2   Frame 3

   a) Copy and complete this table.
      Record data for the first 6 frames.
   b) Determine the first differences.
      Is the relationship between frame number and number
      of tiles linear? Explain.
   c) Graph *Number of tiles* against *Frame number*.
      Does the graph support your answer to part b? Explain.
   d) How many tiles will be in the 9th frame?
      How did you find out?

Frame number	Number of tiles

First differences can be negative.

Example

This table shows the money in a bank account when it was opened, and at the end of each following week.

Number of weeks	Account balance ($)
0	200
1	175
2	150
3	125
4	100

a) Determine the first differences. What do the first differences represent?
b) Is this a linear or non-linear relation? Explain.
c) Graph this relationship. Describe how the patterns in the table are shown in the graph.
d) How much money is in the account after 6 weeks?

Solution

a)

Number of weeks	Account balance ($)	First differences
0	200	
1	175	$175 - 200 = -25$
2	150	$150 - 175 = -25$
3	125	$125 - 150 = -25$
4	100	$100 - 125 = -25$

The first differences represent the change in the account balance each week. Since the first differences are negative, money is being withdrawn from the account.

b) The first differences are equal. The relationship is linear.

c)

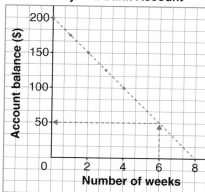

The points lie on a straight line that goes down to the right. For every unit you move to the right, you move 25 units down.

d) From the graph in part c, there are $50 in the account after 6 weeks.

6.1 Recognizing Linear Relations

5. A barrel contained 42 L of water. The water was leaking out. The table shows how the volume of water in the barrel changed every hour.

Time (h)	Volume (L)
0	42
1	38
2	34
3	30
4	26

  a) Determine the first differences.
     What do the first differences represent?
  b) Is the relationship linear or non-linear? Explain.
  c) Graph the relation. Does the graph support your answer to part b? Explain.
  d) How much water would be in the barrel after 6 h? What assumptions did you make?
  e) Write a question you could answer using the data. Answer the question.

6. Assessment Focus The table shows the areas of rectangles for which the length is 3 times the width.

Width (cm)	Area (cm²)
1	3
2	12
3	27
4	48

  a) Sketch the rectangles.
  b) Determine the first differences.
  c) Is the relationship linear or non-linear? Explain.
  d) Graph the relationship. Describe the graph.
  e) Predict the area of the next rectangle in the pattern. Sketch the rectangle. Calculate its area to check your prediction.

7. Laura is swimming lengths to prepare for a triathlon. The table shows the distances she swims in metres.

Number of lengths	Distance (m)
0	0
5	125
10	250
15	375
20	500

  a) By how much are the numbers in the first column increasing?
  b) Determine the first differences. What do the first differences represent?
  c) Is the relationship linear or non-linear? Explain.
  d) Graph the relationship. Does the graph support your answer to part c? Explain.
  e) What is the length of the pool? How did you find out?

8. Take It Further Make your own pattern using square tiles, grid paper, or square dot paper. Choose 2 properties of the figures in your pattern; such as frame number, height, side length, perimeter, area, or number of squares. Investigate whether the number pattern that relates the properties is linear or non-linear. Show your work.

In Your Own Words

Describe two ways to identify whether a relationship is linear or non-linear. Give an example of each way.

6.3 Other Rates of Change

Before an operation, an animal must be "put out" so that it does not feel any pain. Many veterinarians use a drug called Pentothal.

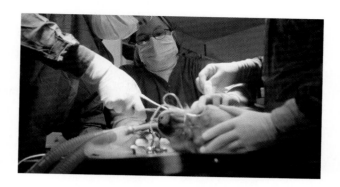

Investigate — Using Tables and Graphs to Explore Rates of Change

The graph shows the amount of Pentothal required for dogs of different masses.

Pentothal Doses for Dogs

Points shown on graph: (1, 23) and (2, 31)

Mass (kg)	Dose of Pentothal (mg)
0	
1	
2	

Record the data from the graph in a table. Include masses up to 6 kg.

Add a third column to the table to show the first differences in the drug dose.

➤ What do you notice about the first differences? Explain.

➤ Select 2 points on the graph. Determine the rate of change. What does this rate of change represent?

➤ Compare the first differences and the rate of change. What do you notice?

Reflect

➤ Is this a linear relation? How do you know?

➤ Explain how you could determine the drug dose for any mass. What assumptions did you make?

Connect the Ideas

The graph shows the average fuel efficiency for an older sport utility vehicle (SUV) and a new Smart car.

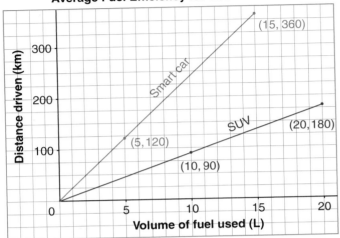

The rate of change can be found for each vehicle.

For the Smart car:
Choose any two points on the line: (5, 120) and (15, 360)
The rise is: 360 km − 120 km = 240 km
The run is: 15 L − 5 L = 10 L
Rate of change = $\frac{\text{rise}}{\text{run}}$
$= \frac{240 \text{ km}}{10 \text{ L}}$
$= 24$ km/L
The Smart car drives 24 km for every litre of fuel it uses.

For the sport utility vehicle:
Choose any two points on the line: (10, 90) and (20, 180)
The rise is: 180 km − 90 km = 90 km
The run is: 20 L − 10 L = 10 L
Rate of change = $\frac{\text{rise}}{\text{run}}$
$= \frac{90 \text{ km}}{10 \text{ L}}$
$= 9$ km/L
The sport utility vehicle drives 9 km for every litre of fuel it uses.

Both rates of change are positive.
The more efficient car has a greater rate of change and a steeper line on the graph. It travels farther for each litre used.

Practice

1. Determine each rate of change.

a) Gas Consumption for a Car

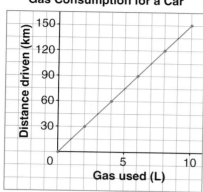

b) Catering Costs for Dinner Guests

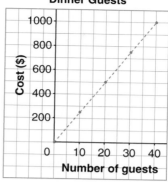

2. What does each rate of change in question 1 represent?

3. Use the data in *Connect the Ideas*, page 202.
The Smart car's gas tank holds 33 L of fuel.
The sport utility vehicle's gas tank holds 80 L.
Determine how far each vehicle could travel on a full tank of gas.

4. Sami conducted an experiment.
He measured the length of a spring when different masses were placed on one end of it.
Sami then calculated how much the spring stretched each time.
These are his results:

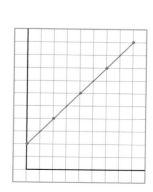

Mass (g)	0	5	10	15	20	25
Stretch (cm)	0	2	4	6	8	10

a) Graph *Stretch* against *Mass*.
b) Does the graph represent a linear relation? Explain.
c) Determine the rate of change of the stretch of the spring with mass.
What does the rate of change represent?

5. **Assessment Focus** Copy this graph on grid paper.
a) Name two quantities that could be represented by this graph.
b) Label each axis.
Give the graph a title.
c) Determine the rate of change.
Explain what it means.

A rate of change can be negative.

Example The graph shows how the height of a small plane changes with time.
a) What is the plane's initial height?
b) Determine the rate of change.
c) What does the rate of change tell you about the plane's flight?

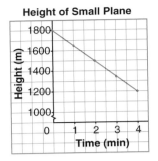

Solution a) The plane's initial height is 1800 m.
This is the height when the time is 0.
b) Choose any two points on the line:
(0, 1800) and (4, 1200)
The rise is: 1200 m − 1800 m = −600 m
The run is: 4 min − 0 min = 4 min
$$\frac{\text{rise}}{\text{run}} = \frac{-600 \text{ m}}{4 \text{ min}}$$
$$= -150 \text{ m/min}$$
The rate of change is −150 m/min.
c) The height is decreasing at an average rate of 150 m per minute. The plane is descending.

6. Suppose the plane in the *Guided Example* continues to descend at the same rate. How many minutes will it take until the plane lands? Explain your answer.

7. The Jasper Tramway is a cable car in the Canadian Rockies. The graph shows how the cable car's height changes on a 7-min trip.
a) What are the initial and final heights, to the nearest 100 m?
b) Determine the approximate rate of change.
c) What does the rate of change tell you about the cable car's trip?

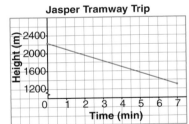

8. **Take It Further** The graph shows the costs of a skating party.
a) Determine the rate of change for each portion of the graph.
b) What do the different rates of change tell you about the costs?
c) Suppose you have a skating party for 8 people. How much will you pay for each person?

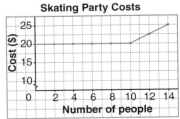

In Your Own Words

Give an example of a rate of change that is not an average speed. Describe what the rate of change measures. Include a graph.

6.4 Direct Variation

Local councils send out leaflets with their water bills. These leaflets describe how water is wasted.

Investigate — Interpreting a Graph that Passes Through the Origin

A faucet leaks. The volume of water wasted and the times when the volume was measured are shown.

➤ What patterns do you see in the table?

➤ Graph the data. Plot *Volume of water wasted* against *Time*. Describe the graph.

➤ Choose two points on the graph. Determine the rate of change.

➤ Repeat the rate of change calculation using two different points on the line. What do you notice? Explain what the rate of change represents.

Time (s)	Volume of water wasted (mL)
0	0
10	15
20	30
30	45
40	60
50	75

➤ Determine the first differences. How are they related to the rate of change? How are the first differences related to the change in times?

➤ How could you determine the volume of water wasted after each time: 100 s? 15 s? 1 s? 22 s?

➤ Suppose you know the time since measuring began. How can you determine the volume of water wasted? Write a rule to determine this volume.

Reflect

➤ Suppose you know the volume of water wasted in 20 min. How could you determine the volume wasted in 40 min?

➤ About how much water would be wasted by the leaking faucet in one day? Explain how you know.

Connect the Ideas

Rhonda bought sesame snacks at the bulk food store.
The cost was $1.10 per 100 g.
Here is a table of costs for different masses.

When the mass of sesame snacks doubles, the cost doubles.
When the mass of sesame snacks halves, the cost halves.

	×2		÷2	
Mass of sesame snacks (g)	200	400	600	800
Cost ($)	2.20	4.40	6.60	8.80
	×2		÷2	

Graph the data

The graph is a straight line that passes through the origin (0, 0).
This illustrates **direct variation**.
We say that the cost *varies directly* as the mass.
When two quantities vary directly, they are *proportional*.
The cost is *directly proportional* to the mass.

Recall the work in Chapter 4 on proportional reasoning. The cost of sesame snacks is directly proportional to their mass.

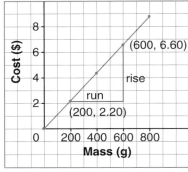

Cost of Sesame Snacks

Determine the rate of change

Choose any two points on the graph: (200, 2.20) and (600, 6.60)
The rise is: $6.60 − $2.20 = $4.40
The run is: 600 g − 200 g = 400 g

$$\frac{\text{rise}}{\text{run}} = \frac{\$4.40}{400 \text{ g}}$$
$$= \$0.011/\text{g}$$

The rate of change is $0.011/g.
The rate of change is the unit price of sesame snacks.

Write a rule

Cost of sesame snacks in $ = (0.011) × (mass of snacks)

↑ the rate of change in dollars per gram   ↑ the mass in grams

We can use the rule to determine how much Rhonda has to pay for 350 g of sesame snacks.
Substitute 350 for the mass of snacks.
Cost of sesame snacks in $ = 0.011 × 350
= 3.85
The cost of 350 g of sesame snacks is $3.85.

Practice

1. Does each graph represent direct variation? Explain how you know.

a)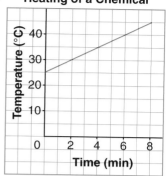
Heating of a Chemical

b)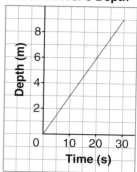
Scuba Diver's Depth

c)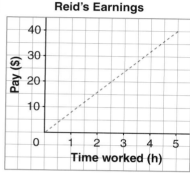
Reid's Earnings

d)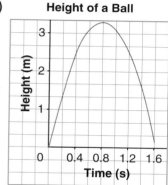
Height of a Ball

2. The perimeter of a square varies directly as its side length.

Side length (cm)	Perimeter (cm)

a) Complete a table to show the perimeters of squares with side lengths from 0 cm and 5 cm.
b) Graph the data. Determine the rate of change.
c) Write a rule to determine the perimeter when you know the side length.
d) What is the perimeter when the side length is 11 cm?

3. About one-third of all water used in homes is for toilet flushes. The table shows water use for a standard toilet manufactured after 1985.

Number of flushes	5	10	15	20	25	30
Water use (L)	65	130	195	260	325	390

a) Graph the data.
   Is this a direct variation situation? Explain.
b) Determine the rate of change.
   What does the rate of change represent?
c) Write a rule to determine the water use when you know the number of flushes.
d) How much water is used for 17 flushes?

6.4 Direct Variation **207**

4. "Nature's Best" sells bird seed for $0.86 per kilogram.
   a) Copy and complete this table for masses from 1 kg to 6 kg.
   b) Graph the data.
      Does the graph represent direct variation? Explain.
   c) Write a rule to determine the cost when you know the mass of seed.
   d) How much would it cost to buy 13 kg of seed?
   e) Will the cost for 26 kg be twice the cost for 13 kg? Explain your thinking.
      To check your answer, use your rule to determine the cost for 26 kg.

Mass of bird seed (kg)	Cost ($)

You can use an equation to calculate values.

Example

Gas for a car is sold by the litre. Here are the costs of gas for 5 customers at a gas station.

Volume of gas (L)	Cost ($)
10	8.50
16	13.60
6	5.10
24	20.40
18	15.30

a) Graph the data.
   Does the graph represent direct variation? Explain.
b) Determine the rate of change.
   Explain what it represents.
c) Write an equation for this relation.
d) Use the equation to determine the cost of 25.8 L of gas.

Solution

a)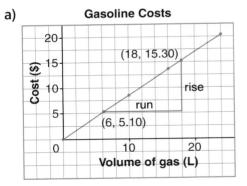

Yes, this graph represents direct variation.
The graph is linear and passes through the origin.

b) From the graph
The rise is: $15.30 − $5.10 = $10.20
The run is: 18 L − 6 L = 12 L

The rate of change is: $\frac{\text{rise}}{\text{run}} = \frac{\$10.20}{12\ L}$
$= \$0.85/L$

The rate of change is the change in price for each litre of gas.
So, the rate of change is the unit cost of gas; that is, the cost for 1 L.

c) The cost of buying gas can be described by the rule:
   Cost of gas in $ = (unit price in $/litre) × (volume of gas in litres)
   Let C represent the cost in dollars.
   The unit price is $0.85/L.
   Let v represent the volume of gas in litres.
   Then an equation is: $C = 0.85 \times v$
   or, $C = 0.85v$

d) Use the equation: $C = 0.85v$
   To determine the cost of 25.8 L of gas, substitute: $v = 25.8$
   $C = 0.85 \times 25.8$
   $= 21.93$
   25.8 L of gas cost $21.93.

5. Kaitlin works planting trees in Northern Ontario. She is paid 16¢ for each tree she plants.
   a) Complete a table to show how much Kaitlin would earn for planting up to 1000 trees.
   b) Graph Kaitlin's earnings.
      Does the graph represent direct variation? Explain.
   c) What is the rate of change?
      What does it represent?
   d) Write an equation for this relationship.
   e) An experienced planter can plant between 1000 and 3000 trees a day.
      Use the equation to determine how much Kaitlin would earn when she plants 1700 trees in one day.

Number of trees, n	Earnings, E ($)
0	0
100	

6. Jamal measured several circles.
   He used thread to measure the circumference and a ruler to measure the diameter.
   Here are his results.
   a) Graph the data.
      Does the graph represent direct variation? Explain.
   b) Determine the rate of change.
      Explain what it represents.
   c) Write an equation for this relationship.
   d) Use the equation to estimate the circumference of a circle with diameter 13 cm.

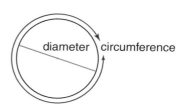

Diameter, d (cm)	Circumference, C (cm)
0	0
12	37.2
2	6.2
7	21.7
5	15.5

7. Drew filled three bags of trail mix at the bulk store.

   a) Graph *Cost* against *Mass* for the trail mix.
      Does the graph represent direct variation? Explain.
   b) Determine the rate of change. Explain what it represents.
   c) Write an equation for this relation.
      Use m for the mass of trail mix in grams and C for the cost in dollars.
   d) Determine the cost of 340 g of trail mix. How did you do this?
   e) Write a question you could answer using the graph or equation.
      Answer the question.

8. **Assessment Focus** Look back at the situations and the equations that represent them. Use these to help you with this question.
   Describe a situation that could be modelled by each equation.
   a) $C = 0.90v$
   b) $d = 90t$
   c) $C = 5.50n$

9. Find a flyer from a grocery store or bulk food store that lists prices per 100 g.
   Choose an item from the flyer.
   Write a question about the item.
   Follow the style of question 7.
   Prepare a sample solution for your question.
   Exchange questions with a classmate.
   Answer the question you receive.
   Check each other's work.

10. **Take It Further** Anastasia drives 55 km to work.
    One day, she left home at 7:00 a.m. and arrived at work at 8:15 a.m.
    a) What would a graph of distance against time look like for Anastasia's drive to work?
       Do you have enough information? Explain.
    b) What is Anastasia's average speed? Explain.

In Your Own Words

How do you know when a graph represents direct variation?
Explain with examples of graphs and equations.

6.5 Partial Variation

Carlo works for a car-sharing co-operative.
The co-op owns a few cars.
Co-op members can arrange to use the cars by the hour, day, or weekend.

Investigate — A Linear Relation Involving a Fixed Cost

For car sharing:

Total cost = fixed cost + cost depending on distance driven

The co-op pays for gas and other expenses.

Carlo prepared a table to show the total cost to use a car for a day.
Some parts of the table are blank because coffee was spilled on it.

Work with a partner to complete the table.

Car Co-op Costs	
Distance driven (km)	Total cost ($)
0	
10	
20	
30	
40	
50	
60	49.00
70	50.50
80	52.00
90	53.50
100	55.00

➤ What patterns do you see in the table?

➤ What is the fixed cost?
   What is the cost per kilometre?

➤ Graph the data.

➤ Write a rule to calculate the total cost when you know the distance driven.

➤ Write an equation. Use C for the total cost in dollars and n for the distance driven in kilometres.

Reflect

Compare your completed table with that of another pair of students.

➤ What strategies did you use to complete the table?

➤ How do the patterns in the table help you determine the fixed cost? The cost per kilometre?

➤ Use your equation to check the cost when a car is driven 90 km.

Connect the Ideas

The cost of a pizza with tomato sauce and cheese is $9.00.
It costs $0.75 for each additional topping.
This table shows the cost of a pizza with up to 8 additional toppings.

Make a table

Toppings	Cost ($)
0	9.00
2	10.50
4	12.00
6	13.50
8	15.00

The cost, in dollars, for a 4-topping pizza is $9 + (4 \times 0.75) = 12.00$

Graph the data

The points are joined with a broken line since we cannot order a fraction of toppings.

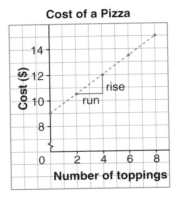

All points lie on a straight line.
The graph does *not* pass through the origin.

This illustrates **partial variation**.
The cost of a pizza is the sum of a fixed cost and a variable cost.

The point where the line crosses the vertical axis is the **vertical intercept**.
On this graph, the vertical intercept is $9.00.
It represents the cost of a pizza with 0 extra toppings.

Determine the rate of change

The rise is: $12.00 − $10.50 = $1.50
The run is: 4 toppings − 2 toppings = 2 toppings
The rate of change is: $\frac{\text{rise}}{\text{run}} = \frac{\$1.50}{2 \text{ toppings}}$
$= \$0.75/\text{topping}$

The rate of change is equal to the cost per topping.

Write a rule

Cost of pizza in $ = 9.00 + (0.75 × number of toppings)

This is the fixed cost in dollars.

This is the variable cost in dollars because it depends on the number of toppings.

Write an equation

Let C represent the cost in dollars.
Let n represent the number of toppings.
Then, $C = 9.00 + 0.75n$

vertical intercept   rate of change

Practice

1. Which graphs represent partial variation? How do you know?

a)

b)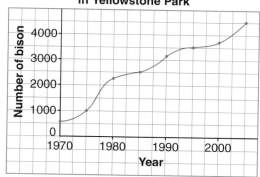

c)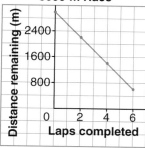

d)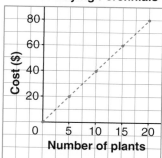

2. For each graph in question 1 that represents partial variation, determine the vertical intercept and the rate of change.

6.5 Partial Variation **213**

3. When it is windy, it feels colder than the actual temperature. The table shows how cold it feels at different temperatures when the wind speed is 10 km/h.

Actual temperature (°C)	−20	−15	−10	−5	0	5
Wind chill equivalent temperature (°C)	−27	−21	−15	−9	−3	3

a) Graph *Wind chill equivalent temperature* against *Actual temperature*.
b) Does this situation represent partial variation? Explain.
c) Use the graph.
About how cold does it feel when there is a 10-km/h wind and the actual temperature is −14°C? 3°C?

4. The temperature at which water boils depends on the altitude; that is, the height above sea level. The table shows how the two quantities are related.

Altitude (m)	0	1200	2400	3600	4800	6000
Boiling point (°C)	100	96	92	88	84	80

a) Graph *Boiling point* against *Altitude*. Did you join the points with a broken line or a solid line? Explain.
b) Does this situation represent partial variation? Explain.
c) Every year, tourists travel to Tanzania to climb Mount Kilimanjaro. Trekkers must boil their drinking water. Use the graph to estimate the boiling point of water at each camp:
  i) Machame camp, 3100 m
  ii) Barafu camp, 4600 m

5. A school is planning an athletic banquet. The banquet hall charges a fixed cost of $500, plus $25 per person for the food.
a) Make a table to show the total cost for up to 50 guests.
b) Graph *Cost* against *Number of guests*. Did you join the points with a broken line or a solid line? Explain.
c) Does the graph represent a partial variation? How do you know?
d) What is the rate of change? What does it represent?
e) How much would the banquet cost for 45 guests? 90 guests? How do you know?
f) Is the cost for 45 guests one-half the cost for 90 guests? Explain.

You have determined values by using a graph.
This example shows how to determine values using an equation.

Example

Jolanda has a window cleaning service.
She charges a $12 fixed cost, plus $1.50 per window.
a) Write an equation to determine the cost when the number of windows cleaned is known.
b) What does Jolanda charge to clean 11 windows?

Solution

a) The cost is $12.00 plus $1.50 × the number of windows.
Let C represent the cost in dollars.
Let n represent the number of windows.
The equation is: $C = 12.00 + (1.50 \times n)$
or, $C = 12.00 + 1.50n$

b) Use the equation: $C = 12.00 + 1.50n$
To determine the cost for 11 windows, substitute: $n = 11$
$C = 12.00 + 1.50 \times 11$
$= 12.00 + 16.50$
$= 28.50$

Use the order of operations.

Jolanda charges $28.50 to clean 11 windows.

6. A phone company charges $20 per month, plus $0.10 per minute for long distance calling.
   a) Write an equation to determine the total monthly charge, C dollars, for n minutes of long distance calling.
   b) Determine the total cost for 180 min of long distance calls one month.

7. Assessment Focus Nuri is growing his hair to donate to a charity.
The charity makes wigs for children with cancer.
Nuri's hair is now 25 cm long. It grows 1.3 cm every month.
   a) Write an equation to determine Nuri's total hair length, h centimetres, after n months.
   b) How long is his hair after 6 months? 9.5 months? How do you know?
   c) Write a question that can be answered using the equation.
      Trade questions with a classmate. Answer the question you receive.

8. Look back at the situations and the equations that represent them.
Use these to help you with this question.
   a) Describe a situation that could be modelled by each equation.
      i) $C = 150 + 20n$　　ii) $\ell = 20w$　　iii) $C = 0.75m$
   b) Which equations in part a represent direct variation? Partial variation? How do you know?

6.5 Partial Variation

9. A school theatre group has posters printed for a performance.
   The cost to print a poster is the sum of a fixed cost and a variable cost.
   It costs $75 to print 100 posters and $150 to print 400 posters.
   a) Graph *Cost* against *Number of posters*.
   b) What is the fixed cost? How can you determine this from the graph?
   c) What is the rate of change? What does it represent?
   d) Write an equation that describes this linear relation.
   e) How much does it cost to print 250 posters? How did you find out?

10. A taxi company charges a fixed cost, called the initial meter fare, and a cost per kilometre.
    A 5-km trip costs $11.25 and a 10-km trip costs $19.75.
    a) Graph *Cost* against *Distance*.
    b) What is the fixed cost?
       How can you determine this from the graph?
    c) What is the rate of change? What does it represent?
    d) Write an equation that describes this linear relation.
    e) How much does it cost to ride 15 km?
    f) This company offers a flat rate fare of $90 for travel from Waterloo to Lester B. Pearson International Airport.
       The airport is 96 km from Waterloo.
       Which is cheaper: the flat rate fare or the partial variation fare?
       How do you know?

11. **Take It Further** In a scavenger hunt, players have 1 h to find 50 items from a list.
    Players are awarded 5 points for every item handed in.
    Players who hand in their items within 45 min are awarded a 20-point bonus.
    a) Suppose a player finishes within 45 min.
       i) Is her score an example of partial variation? Explain.
       ii) Write an equation to determine the total number of points when the number of items is known.
    b) Yvonne finishes within 45 min. She scores 220 points.
       Determine how many items Kevin needs to find to beat Yvonne in each case.
       i) Kevin also finishes within 45 min.
       ii) Kevin takes the full hour.

In Your Own Words

What is partial variation?
Use an example and a graph to explain.

Using a Graphing Calculator to Graph an Equation and Generate a Table of Values

You will need a TI-83 or TI-84 graphing calculator.

Consider the *Guided Example* in Section 6.5, page 215.
Here is the equation for Jolanda's window cleaning service:
C = 12.00 + 1.50n
where C dollars is the cost and n is the number of windows

The graphing calculator uses X to represent the variable on the horizontal axis and Y to represent the variable on the vertical axis.
To match this style, the cost equation is written as:
Y = 12.00 + 1.50X

Follow these steps to graph the equation and generate a table of values.

To make sure there are no equations in the [Y=] list: Press [Y=]. Move the cursor to any equation, then press [CLEAR]. When all equations have been cleared, position the cursor to the right of the equals sign for Y1. To enter the equation, press: [1][2][.][0][0][+][1][.][5][0][X,T,Θ,n]	Plot1 Plot2 Plot3 \Y1■12.00+1.50X \Y2= \Y3= \Y4= \Y5= \Y6= \Y7=
To adjust the window settings to display the graph: Press [WINDOW]. Set the options as shown. To set an option, clear what is there, enter the value you want, then press [ENTER]. Use the ⊙ key to scroll down the list.	WINDOW Xmin=0 Xmax=10 Xscl=1 Ymin=0 Ymax=30 Yscl=1 Xres=1■

Since X represents the number of windows washed and Y represents the cost, we are only interested in positive values of X and Y.

Press [GRAPH]. A graph like the one shown here should be displayed. The points lie on a line. Since X is a whole number, the graph should be a series of dots joined with a broken line. However, the calculator joins the points with a solid line.	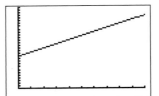
To display a table of values: Press [2nd] [WINDOW] to display the table setup screen. Press [0] [ENTER]. This sets the initial value of X in the table to 0. Press [1] [ENTER]. This increases the value of X by 1 each row. Press [ENTER] ⊙ [ENTER]. Your screen should look like the one shown here.	
Press [2nd] [GRAPH] to display the table. Only the first 7 rows are displayed. Use the ⊙ key to scroll down the table. Confirm that when 11 windows are washed, the cost is $28.50.	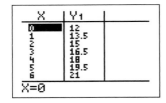

In Section 6.5, question 8, page 215, you described situations that could be modelled by each of 3 equations.
Complete these steps for each equation.

Use pages 217 and 218 as a guide.

➢ Use a TI-83 or TI-84 calculator to graph the equation, using appropriate window settings.

➢ Display a table of values for the graph.

➢ Choose a value of X that makes sense for the situation you described in question 8.
Use the table to determine the corresponding value of Y.
Describe what the values represent.

The 25-m Sprint

In track and field, the 60-m sprint is the shortest race. In this game, you will be racing to 25 m. But you will not be running. Your challenge is to graph a linear relation that reaches 25 m as quickly as possible.

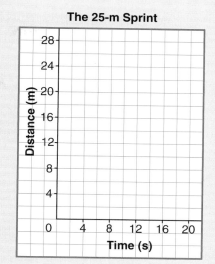

The 25-m Sprint

Play with a partner.
You will need 3 number cubes, a ruler, several copies of the grid shown at the right, and a different coloured pencil for each player.
The first player rolls the number cubes and records the numbers shown.
The player chooses:
- one number to be the vertical intercept
- another number to be the rise
- the third number to be the run

The first player graphs her relation:
Plot the vertical intercept.
Use the rise and run to plot the next point.
Plot several points.
Use a ruler to draw a line through them.
Extend the line until it passes beyond 25 m.

Play passes to the second player.
This player rolls the number cubes and decides what the numbers will represent.
The player follows the steps above to graph his relation on the same grid, using a different colour.

Compare the 2 lines.
Whose line reached 25 m first?
This person is the winner.

Play the game several times.
Think about the strategies you use when you decide what the numbers rolled will represent. Describe a winning strategy.
Test your ideas by playing the game again.

> Remember that the rise is the vertical distance between two points and the run is the horizontal distance between these points.

Mid-Chapter Review

6.1, 6.3

1. The table shows how the height of a small plane changes with time.

Time (min)	Height (m)
0	0
1	200
2	400
3	600
4	800

a) Determine the first differences. What do they represent?
b) Is the relationship linear? Explain.
c) Graph *Height* against *Time*. Determine the rate of change. How does it compare with the first differences?
d) At this rate, how many minutes will it take the plane to reach its cruising height of 1600 m? Explain how you found your answer.

6.2

2. The fastest elevator in the world is in the Taipei 101 office tower, Taiwan. The tower is about 510 m high. It is the tallest building in the world. It takes the elevator about 30 s to reach the top.
a) Graph *Height* against *Time* for the elevator.
b) Determine the height after each time.
   i) 15 s   ii) 20 s
   Which method did you use to find the heights?
c) What is the rate of change of distance with time?
d) What is the average speed for the trip?

6.4, 6.5

3. The cost of renting a kayak is $33 per day.
a) Copy and complete this table to show the cost to rent a kayak for up to 10 days.

Rental time (days)	Cost ($)

b) Graph *Cost* against *Rental time*.
c) Does this graph represent direct variation or partial variation? Explain how you know.
d) Determine the rate of change. What does it represent?

6.5

4. In a cookbook, the time to cook a turkey is given as:
30 min per kilogram of turkey, plus an additional 15 min
a) Complete a table like the one below for turkeys with masses up to 10 kg.

Mass of turkey (kg)	Cooking time (min)
0	15
2	
4	

b) Graph *Cooking time* against *Mass*. Use the graph to determine the time to cook a 5-kg turkey.
c) Write a rule for the cooking time.
d) Write an equation for the cooking time, t minutes, for a turkey with mass m kilograms.
e) Determine the cooking time for a 7.2-kg turkey. How many different ways could you find this time? Explain.

CHAPTER 6: Linear Relations

6.6 Changing Direct and Partial Variation Situations

Investigate — Making Changes to Linear Relations

Your teacher will give you a copy of the graphs at the left.

➤ Determine whether each graph represents direct variation, partial variation, or neither. Give reasons for your choices.

➤ Match each graph with its description below.
Label your copy of the graphs with as much detail as possible.
• The "Creature Comforts" dog kennel costs $28 per day.
• Rana rents a moped while on vacation.
She is charged an initial fee of $35 plus $2 for each litre of fuel consumed.

➤ Write an equation for each graph.

➤ Suppose the daily cost at the dog kennel increases to $30.
How does the graph change?
Sketch the new graph on the grid.
How does the equation change?
Write the new equation.

➤ When Rana rented a moped the next year, the prices had changed.
The initial fee had been reduced to $28.
The price for fuel had increased to $3 per litre.
How does the graph change?
Sketch the new graph on the grid.
How does the equation change?
Write the new equation.

Reflect

Compare your answers with those of other students.

➤ How are graphs of direct variation and partial variation alike? How are they different?

➤ How did the changes in each situation affect the graph?

➤ How do these changes affect the equation?

Connect the Ideas

Ben works part time selling newspaper ads.
He is paid $18 for each ad he sells.
His pay varies directly as the number of ads sold.
The rate of change is $18/ad.
The rule is:
Pay in dollars = 18 × number of ads sold
Let P represent the pay in dollars.
Let n represent the number of ads sold.
Then the equation is:
$P = 18n$

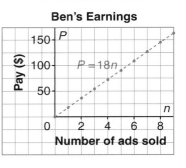

Ben is given a pay raise.
He is now paid $22 for each ad he sells.
The rate of change is now $22/ad.
Here is the new graph and the original graph.
The rule for the new graph is:
Pay in dollars = 22 × number of ads sold
The new equation is:
$P = 22n$

Both graphs pass through the origin and go up to the right.
The purple graph has the greater rate of change.
The purple graph is steeper than the red graph.
Beyond the origin, the purple graph lies above the red graph.

Practice

1. Does each situation represent direct variation or partial variation? Explain how you know.
   a) Lily is paid $5 per hour for raking leaves.
   b) The printing of brochures costs $250, plus $1.25 per brochure.
   c) Jordan is paid $30 per day, plus $2.00 for every magazine subscription he sells.

2. Does each equation represent direct variation or partial variation? Explain how you know.
   a) $C = 4n + 30$
   b) $P = 4s$
   c) $d = 65t$
   d) $d = 400 - 85t$

3. The cost for Best Cellular cell phone plan is $0.20 per minute.
   a) Does the cost represent direct variation or partial variation? How do you know?
   b) Make a table for the costs. Use intervals of 10 min from 0 min to 100 min.
   c) Graph the data.
   d) Write an equation to determine the cost, C dollars, for t minutes of calls.
   e) The company increases its fees to $0.25 per minute.
      i) How does the graph change? Draw the new graph on the grid in part c.
      ii) How does the equation change? Write the new equation.

Changes can be made to a situation described by a partial variation.

Example

Jocelyn works full time selling newspaper ads. She is paid $32 a day, plus $8 for every ad she sells.
a) Make a table of Jocelyn's earnings for daily sales up to 60 ads.
b) Graph the data.
c) Write an equation to determine Jocelyn's earnings, E dollars, when she sells n ads.
d) There is a change in management.
   Jocelyn is now paid $30 per day and $10 for every ad she sells.
   i) Graph Jocelyn's earnings for daily sales up to 60 ads.
      How is the graph different from the graph in part b?
   ii) Write the new equation for Jocelyn's earnings.

Solution

a)

Number of ads	Earnings ($)
0	32
20	$32 + (8 \times 20) = 192$
40	$32 + (8 \times 40) = 352$
60	$32 + (8 \times 60) = 512$

b)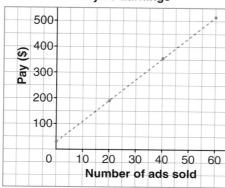

The vertical intercept is $32.
The rate of change is $8/ad.

c) The rule for Jocelyn's earnings is:
   Earnings in dollars = $32 + (8 \times$ number of ads sold$)$
   An equation is: $E = 32 + 8n$

d) i) Here is the graph with the new rate. The vertical intercept is $30.
The rate of change is $10/ad.
ii) The new equation is: $E = 30 + 10n$

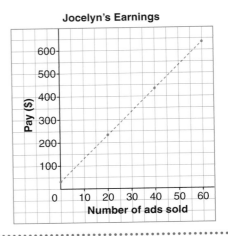

4. Use the data in the *Guided Example*. Which pay scale was better for Jocelyn? Give reasons for your answer.

5. The Best Cellular cell phone company has another plan for cell phone users. This plan costs $12 per month, plus $0.10 per minute.
  a) Make a table of monthly costs for calling times from 0 min to 100 min.
  b) Graph the data.
  c) Write an equation to determine the monthly cost, C dollars, for t minutes of calls.
  d) The company increases its fixed monthly fee to $20 but decreases its cost per minute to $0.08.
    i) How does the graph change? Draw the new graph on the grid in part b.
    ii) Write the new equation.

6. Assessment Focus Use the data and your results from questions 3 and 5. For each plan, does the cost double when the time doubles? Explain.

7. Take It Further A car rental agency charges a daily rate of $29. If the distance driven is more than 100 km, a charge of $0.10 per kilometre is added for each kilometre over 100 km.
  a) Make a table to show the cost every 50 km up to 300 km.
  b) Graph the data.
  c) Does the graph represent partial variation, direct variation, or neither? How do you know?
  d) Describe the graph. Explain its shape.
  e) How much would you have to pay if you drove 35 km? 175 km? 280 km?

In Your Own Words

How are the graph and equation of a partial variation situation different from those of direct variation? Use examples in your explanation.

Using CAS to Explore Solving Equations

You will need a TI-89 calculator.

Step 1

- Clear the home screen:
  [HOME] [F1] 8 [CLEAR]
- Enter the equation $x = 3$:
  [X] [=] 3 [ENTER]
- Multiply each side of the equation by 2:
  [×] 2 [ENTER]
- Add 1 to each side of the equation:
  [+] 1 [ENTER]

At the end of Step 1, your screen should look like this:

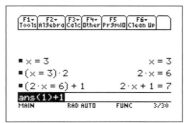

$x = 3$
$2x = 6$
$2x + 1 = 7$ are examples of *equivalent equations*.

Why are the equations equivalent?
How are equivalent equations formed?

Step 2

In Step 1, you started with $x = 3$ and ended with $2x + 1 = 7$.

Suppose you start with $2x + 1 = 7$. What steps would you take to get back to $x = 3$?

Which operations did you use to get from $x = 3$ to $2x + 1 = 7$?

How could you "undo" these operations to get back to $x = 3$?

Step 3

- Clear the home screen:
  [HOME] [F1] 8 [CLEAR]
- Enter the equation $2x + 1 = 7$:
  2[X] [+] 1 [=] 7 [ENTER]
- Subtract 1 from each side:
  [−] 1 [ENTER]
- Divide each side by 2:
  [÷] 2 [ENTER]

At the end of Step 3, your screen should look like this:

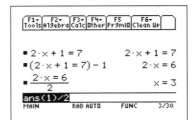

The steps to get back to $x = 3$ are the steps to solve the equation $2x + 1 = 7$.

At each step:
– we are closer to isolating the term containing the variable
– we create a simpler equivalent equation

Technology: Using CAS to Explore Solving Equations

Step 4

- Clear the home screen:
  [HOME] [F1] 8 [CLEAR]
- Enter $2x + 1 = 7$:
  2[X] [+] 1 [=] 7 [ENTER]
- Divide each side by 2:
  [÷] 2 [ENTER]
- Subtract 1 from each side:
  [−] 1 [ENTER]

At the end of Step 4, your screen should look like this:

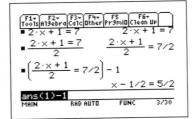

Which extra step is needed to isolate x on the left side of the equation?

What is the result when you isolate x?

Which method, Step 3 or Step 4, is easier to use? Explain.

Step 5

- Work with a partner. Repeat Step 1 with a different start equation. Multiply each side by a different number. Then add a different number to each side, or subtract a different number from each side.
- Trade end equations with your partner.
- Solve your partner's equation.

Which strategies did you use to solve the equations?

Step 6

Solve these equations using paper and pencil.
a) $3x - 1 = 14$
b) $5x + 12 = 2$
c) $15 - 2x = 3$
Use CAS to check your work and to check your solution.

To check if $x = 3$ is a solution to $2x + 1 = 7$, press: 2[X] [+] 1 [=] 7
[|][X] [=] 3 [ENTER].
To test $x = 4$,
press: [▶] [◀] 4 [ENTER]

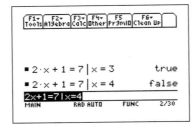

How does the calculator show that $x = 3$ is a solution?

How does it show that $x = 4$ is not a solution?

226   CHAPTER 6: Linear Relations

6.7 Solving Equations

An equation is a statement that two expressions are equal.
Each of these is an equation:
$3x = 21$
$12 = x + 4$
$2x - 7 = 9$

Investigate: Using Inverse Operations to Solve an Equation

Work with a partner.

➢ Start with the equation $x = 2$.
  This is the *start equation*.

➢ Multiply each side of the equation by 3.
  Write the resulting equation.

➢ Add 5 to each side of the equation.
  Write the resulting equation.
  This is the *end equation*.

➢ Suppose you were given the end equation.
  What steps would you take to get back to $x = 2$?

Repeat the steps above with a different start equation.
Multiply each side by a different number.
Then add a different number to each side,
or subtract a different number from each side.
Trade end equations with your partner.
Find your partner's start equation.

Reflect

Share your equations with another pair of classmates.

➢ Find your classmates' start equations.
  What strategies did you use?

➢ How are the operations used to get from the end equation to the start equation related to the operations used to get from the start equation to the end equation?

Connect the Ideas

We can use mental math to solve equations like $3 + x = 5$.
We know that $3 + 2 = 5$, so $x = 2$.

Equations like $4x + 3 = 11$ cannot be easily solved mentally.
Here are two ways to solve the equation.

Use a balance scales model

Let x represent the number of candies in each bag.
There are 4 bags and 3 candies on the left pan.
This is represented by the expression $4x + 3$.
There are 11 candies on the right pan.
This is represented by the number 11.
So, the balance scales model the equation $4x + 3 = 11$.

Each bag contains the same number of candies.

To solve the equation:
Remove 3 candies from each pan.
We now have: $4x = 8$

Divide the candies in the right pan into 4 equal groups.
Each group contains 2 candies.
So, each bag contains 2 candies.
The solution is $x = 2$.

Use computer algebra systems (CAS)

Enter the equation $4x + 3 = 11$.
To solve the equation, use inverse operations.

Isolate the term $4x$ by "undoing" the addition.
The inverse of adding 3 is subtracting 3.
Subtract 3 from each side of the equation.
We now have: $4x = 8$

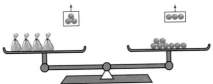

Isolate x by "undoing" the multiplication.
The inverse of multiplying by 4 is dividing by 4.
Divide each side of the equation by 4.
The solution is $x = 2$.

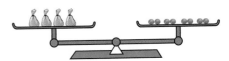

Practice

1. Solve each equation. Explain how you did it.
   a) $3x = 12$
   b) $45 = 5x$
   c) $x - 7 = 11$
   d) $2x = 4$
   e) $4 = 3 + x$
   f) $x + 5 = 14$

2. Write the equation represented by each balance scales. Explain how to solve the equation.

   a)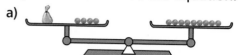
   b)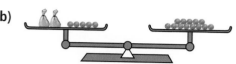
   c)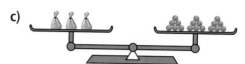
   d)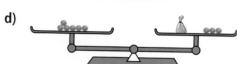
   e)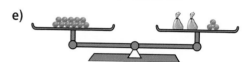
   f)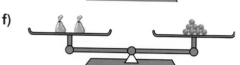

3. Solve each equation.
   a) $3x + 2 = 17$
   b) $2x - 5 = 1$
   c) $33 = 4x + 5$
   d) $7x + 1 = 22$
   e) $2 = 6x - 10$
   f) $5x - 2 = 38$

   Which tools could you use to help you?

   Having trouble? Read Connect the Ideas.

We cannot use balance scales when the solution to an equation is a negative number. Use inverse operations instead.

Example

Solve: $3x + 14 = 2$

Solution Use an inverse operation to isolate $3x$ on the left side of the equation.

$3x + 14 = 2$ Subtract 14 from each side.
$3x + 14 - 14 = 2 - 14$
$3x = -12$
$\frac{3x}{3} = \frac{-12}{3}$ To isolate x, divide each side by 3.
$x = -4$

Check the solution.
Substitute $x = -4$ in $3x + 14 = 2$.
L.S. $= 3(-4) + 14$ R.S. $= 2$
     $= -12 + 14$
     $= 2$

Since the left side equals the right side, $x = -4$ is the correct solution.

4. Solve each equation.
   a) $3x + 4 = 1$
   b) $5 + 2x = 3$
   c) $-3x + 4 = 7$
   d) $2 - 5x = 17$
   e) $7 = 3x + 13$
   f) $-3x - 5 = 4$
   g) $24 = -3x - 6$
   h) $-5 + 4x = -21$

5. Choose 3 equations in question 4. Check your solutions.

6. An equation for the perimeter of a parallelogram is $P = 2b + 2c$.
   a) Determine P when $b = 5$ cm and $c = 7$ cm.
   b) Determine c when $P = 36$ cm and $b = 8$ cm
   c) Determine b when $P = 54$ cm and $c = 12$ cm.

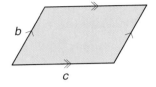

7. **Assessment Focus** In Canada, temperature is measured in degrees Celsius, C. In the United States, temperature is measured in degrees Fahrenheit, F. The equation $9C = 5F - 160$ is used to convert between the two temperature scales.
   a) What is 30°C in degrees Fahrenheit?
   b) What is 14°F in degrees Celsius?
   Explain your work.

8. An empty tanker truck has a mass of 14 000 kg. One barrel of oil has a mass of 180 kg. The equation $M = 14\,000 + 180b$ represents the total mass of the truck, M kilograms, when it contains b barrels of oil. The truck enters a weigh station that shows its total mass is 51 080 kg. How many barrels of oil are on the truck? How do you know?

9. **Take It Further** In each figure:
   a) Write an equation to relate the given measures.
   b) Solve the equation.

   i)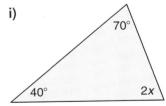
   ii)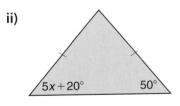
   iii)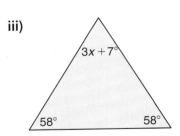

In Your Own Words

Choose one equation you solved in this section.
Explain how you solved it.
Show how to check the solution.

6.8 Determining Values in a Linear Relation

To estimate how much cut lumber a tree can provide, a forester needs to know the size of the tree.
It can be difficult to measure the height of a tree.
It is easier to measure the tree's circumference, then calculate its diameter.
So, it is useful to know how a tree's height is related to its diameter.

In forestry, the diameter of a tree is measured at chest height, 1.37 m above the ground.

Investigate — Determining Values Using a Table, Graph, and Equation

Researchers collected data from more than 100 white spruce trees on Prince Edward Island.
The data were graphed and a line of best fit was drawn.
The equation of the line is: $d = 10h - 50$
where d is the diameter of the tree in centimetres and h is the approximate height of the tree in metres

The relationship between height and diameter changes as trees age. This equation is valid for trees with diameters between 5 cm and 25 cm.

➤ Make a table to show the approximate heights of trees with diameters 5 cm and 25 cm.

Approximate height, h (m)	Diameter, d (cm)
	5
	25

➤ Graph the data.
How could you use the graph to estimate the height of a tree that is not in the table?

➤ How could you use the graph to estimate the diameter of a tree that is not in the table?

> **Reflect**
>
> You have used a table, a graph, and an equation to estimate the heights and diameters of trees.
> Which method do you prefer? Is one method best? Explain.

Connect the Ideas

Janice is a piano teacher.
She earns $23 per hour.
Janice wants to earn $200.
Here are three ways she can determine the time she has to work.

Make a table

Continue the table until the *Earnings* is $200 or greater.

Time worked (h)	1	2	3	4	5	6	7	8	9
Earnings ($)	23	46	69	92	115	138	161	184	207

From the table, Janice has to work 9 h to earn at least $200.

Draw a graph

Janice earns $23/h.
She earns $0 for 0 h worked.
Janice earns $23 × 10 = $230 for 10 h worked.
Since the relationship is a direct variation, plot the points (0, 0) and (10, 230), then draw a broken line through them.
From the graph, Janice has to work about 8.5 h to earn $200.

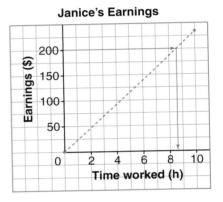

Use an equation

Since the relationship is a direct variation, the rule is:
Earnings in dollars = rate of change × time worked in hours
The rate of change is the hourly rate: $23/h
Let A dollars represent the earnings.
Let n hours represent the time worked.
Then, the equation is: $A = 23n$
To determine how long Janice needs to work to earn $200, substitute $A = 200$ and solve the equation for n.
Use inverse operations.

$200 = 23n$ Divide each side by 23.
$\frac{200}{23} = \frac{23n}{23}$ Use a calculator to determine $200 \div 23$.
$8.696 \doteq n$

Janice has to work about 8.7 h to earn $200.
Since Janice works whole numbers of hours, she has to work 9 h to earn $200.

Practice

1. a) This table shows the cost to rent a canoe. How much would it cost to rent a canoe for each time?

Time (days)	0	1	3	5
Cost ($)	0	24	72	120

   i) 2 days **ii)** 4 days
   How did you find out?

b) This table shows the earnings from ticket sales. How much is earned from the sale of each number of tickets?

Tickets sold	0	20	40	60
Earnings ($)	0	160	320	480

   i) 10 **ii)** 30 **iii)** 50
   How did you find out?

2. These data were recorded for a cheetah chasing prey.

Time (s)	0	5	10	15	20
Distance (m)	0	105	210	315	420

a) Graph the data.
b) Use the graph to determine how long it takes the cheetah to run 400 m.
c) About how long will it take the cheetah to run 500 m? What assumption did you make?
d) Write an equation to determine the distance, d metres, after t seconds.
e) Use the equation to determine how long it will take the cheetah to run 400 m and 500 m. How do your answers compare to those in parts b and c? Which method did you prefer? Explain.

We can also find values for a linear relation that represents partial variation.

Example

A candle is 13 cm long.
It is lit and it burns at the rate of 0.2 cm/min.
a) Write an equation for the height, H centimetres, of the candle after time, t minutes.
b) How long does it take until the candle is 10 cm high?

Solution

a) The rule is:
Candle height = initial height − (rate of change × time)
The range of change is 0.2 cm/min.
So, the equation is:
$H = 13 - (0.2 \times t)$, or
$H = 13 - 0.2t$

b) To determine how long it takes until the candle is 10 cm high, substitute $H = 10$ in the equation, then solve for t.

$$H = 13 - 0.2t$$
$$10 = 13 - 0.2t$$
Use inverse operations.
Subtract 13 from each side.
$$10 - 13 = 13 - 13 - 0.2t$$
$$-3 = -0.2t$$
$$\frac{-3}{-0.2} = \frac{-0.2t}{-0.2}$$
Divide each side by -0.2.
Use a calculator to divide.
$$15 = t$$

It takes 15 min until the candle is 10 cm high.

3. Use the data in the *Guided Example*.
   a) Graph *Height* against *Time*.
   Determine when the candle is 10 cm high. How did you do this?
   b) Make a table of values.
   Use the table to determine when the candle is 10 cm high.
   c) Which of the 3 methods to determine the height of the candle did you prefer? Explain.

4. **Assessment Focus** At the video arcade, Aran bought a game card with a value of $35. Every time he plays a game, $1.40 is deducted from the card.
   a) Make a table for the number of games played.
   Use the table to determine when the card runs out.
   b) Graph the data.
   Use the graph to determine when the card runs out.
   c) Write an equation for the value, V dollars, of the card after n games have been played.
   d) How many games can Aran play before his card runs out? How do you know?
   e) Which of the 3 methods do you prefer? Explain.

Number of games played, n	Value of card, V ($)

5. **Take It Further** Aran could have bought a card that costs $10 more, but only $0.90 is deducted for each game played.
   a) How does this affect the graph?
   b) How many more games could Aran play with this card?
   c) Do you think it is a good deal? Explain.

In Your Own Words

Describe some different ways to determine values in a linear relation. Which way do you prefer? Use an example to explain.

6.9 Solving Problems Involving Linear Relations

Vimy Ridge, in Northern France, was the site of one of the most infamous battles of World War I.
In 2007, groups of Canadian high school students will visit the site to commemorate the 90th anniversary of the battle.
When they travel to France, students will convert their money from dollars to Euros.

Investigate — Developing an Equation from a Graph

The price of one currency in terms of another is called the *exchange rate*.
Exchange rates vary from day to day.
This graph is based on the average exchange rate in January 2006.
Your teacher will give you a copy of the graph.

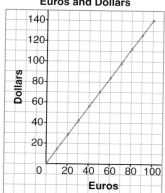

Relationship between Euros and Dollars

➤ Does this graph represent direct variation or partial variation? Explain how you know.

➤ Use the graph.
Determine the value of $70, in Euros.
What is the value of 100 Euros, in dollars?

➤ Determine the rate of change.
What does it represent?

➤ Write an equation for the value, d Canadian dollars, of E Euros.

➤ Use the equation to determine the value of $100, in Euros.
What is the value of 20 Euros, in dollars?
Can you use the graph to determine these values? Explain.

Reflect

➤ What strategies did you use to write the equation of a relationship given its graph?

➤ When might you want to use the equation rather than the graph?

Connect the Ideas

Alexei went on a ski trip that cost $1700.
He borrowed the money from his parents to pay for the trip.
Every month, he pays them back the same amount of money.

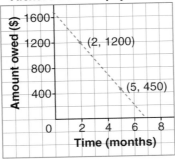

Determine the rate of change

From the graph
After 2 months, Alexei owes $1200.
After 5 months, he owes $450.
So, the rise is: $450 − $1200 = −$750
The run is: 5 months − 2 months = 3 months

Rate of change $= \frac{\text{rise}}{\text{run}}$

$= \frac{-\$750}{3 \text{ months}}$

$= -\$250/\text{month}$

This means that every month, Alexei owes $250 less than the month before.
That is, his payments are $250 each month.

Write a rule

The vertical intercept is $1700.
This represents the amount that Alexei borrowed;
that is, the amount when measurements of time began.

Amount owed = $1700 − ($250/month × time in months)

Write an equation

Let A dollars represent the amount.
Let t months represent the time.
Then, an equation is: $A = 1700 - 250t$

We can find out when Alexei will have paid off his loan.

➤ Use the graph.
At 7 months, the amount owed is approximately zero.
This means that Alexei has paid off the loan.

➤ Use the equation.
The amount Alexei owes is $0.
Substitute $A = 0$, then solve for t.

$A = 1700 - 250t$
$0 = 1700 - 250t$ Add $250t$ to each side.
$250t = 1700 - 250t + 250t$
$250t = 1700$ Divide each side by 250.
$\frac{250t}{250} = \frac{1700}{250}$
$t = 6.8$

Since the number of months is a whole number, it takes Alexei 7 months to pay off his loan.

> When we add $250t$ to each side, then divide each side by 250, we are using inverse operations.

Practice

Your teacher will provide larger copies of the graphs in this section.

1. For each graph, make a table of values.

a) Taxi Fares
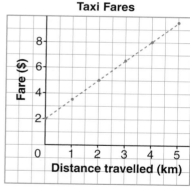

b) Laurie's Loan Repayments
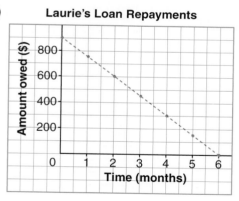

c) Relationship between Mexican Pesos and Dollars
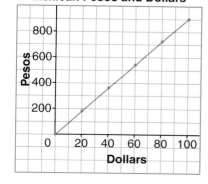

6.9 Solving Problems Involving Linear Relations   **237**

2. Rosa made this pattern with toothpicks. She continued it for 7 frames.

Frame 1

Frame 2

Frame 3

Rosa drew this graph.

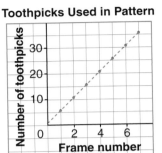

a) Does this graph represent direct variation or partial variation? How do you know?
b) Make a table of values to show how many toothpicks are needed for each frame.
c) Write an equation that relates the number of toothpicks, T, and the frame number, n.
d) How many toothpicks would you need to build the 17th frame in this pattern? Which method did you use to find out?

3. The graph below models the motion of two cars.
a) Use the graph. Copy and complete the table.

Distance Travelled by Two Cars

Time (h)	Car A Distance (km)	Car B Distance (km)
0		
0.5		
1		
1.5		
2		

b) For each car, write an equation that relates distance, d kilometres, and time, t hours.
c) Which car has the greater average speed? How do you know?
d) How far has each car travelled after 1.75 h?
e) How long does it take each car to travel 110 km?

We can graph a relation from its description.

Example

Ashok and Katie each recorded a distance for a ball rolling over a period of time.
Ashok found that the ball rolled 9 m in 3 s.
Katie found that the ball rolled 6 m in 2 s.

a) Draw a distance-time graph for the ball.
   Assume the ball continues at the same average speed.
b) When will the ball have rolled 10 m?
c) Write an equation that relates the distance, d metres, and the time, t seconds.
d) Use the equation.
   How far did the ball roll in 8 s?
e) Compare your solutions in parts b and d.

Solution

a) Complete a table of values.
   Graph the data.

Time, t (s)	0	2	3
Distance, d (m)	0	6	9

The points lie along a straight line that passes through the origin.
Extend the line.

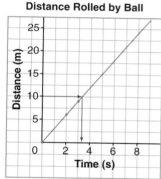

b) To determine when the ball rolled 10 m, use the graph.
   From the graph, the ball reached 10 m after about 3.3 s.

c) The situation is modelled by direct variation.
   Distance in metres = average speed in m/s × time in seconds
   The average speed is the rate of change.
   The ball rolled 9 m in 3 s.

   Average speed = $\frac{\text{distance}}{\text{time}}$
   $= \frac{9 \text{ m}}{3 \text{ s}}$
   $= 3$ m/s

   > The graph will not go on forever. The ball will eventually slow down and stop.

   So, distance in metres = 3 m/s × time in seconds
   The equation is: $d = 3t$

d) To determine how far the ball rolled in 8 s, substitute $t = 8$.
   $d = 3 \times 8$
   $= 24$

6.9 Solving Problems Involving Linear Relations   **239**

e) In part b, from the graph, the answer was approximate.
To determine the answer, the graph was extended to the right.
Drawing the graph helps visualize the relationship between the quantities.
In part d, from the equation, the answer was exact.

4. Liam sells jewellery.
   He earns a basic wage, plus commission.
   Liam earned $550 when he sold $1000 worth of jewellery.
   He earned $750 when he sold $5000 worth of jewellery.
   How much does Liam earn when he sells $6000 worth of jewellery?
   a) Use a table to solve the problem.
   b) Use an equation to solve the problem.
   c) Which method in parts a and b do you prefer? Explain.
   d) How could you have solved the problem a different way? Explain.

5. **Assessment Focus** Rain water is collected in a 350-L barrel.
   During one storm, the volume of water in the barrel after 1 min was 65 L.
   The volume after 5 min was 125 L.
   Suppose water continues to be collected at this rate.
   a) How many minutes will it take for the barrel to fill?
   b) How many different ways could you determine the answer to part a?
      Describe each way.
      Which way do you prefer?

6. Suppose the rate of water collection in question 5 was 12 L per minute.
   How many minutes would it take for the barrel to fill?

7. **Take It Further** Use the table.
   a) Write an equation to show the relationship between the distance, d, and time, t.
   b) Describe a situation that could be modelled by this table and equation.

Time, t (min)	0	1	2	3	4
Distance, d (m)	3	5	7	9	11

In Your Own Words

Describe how to write an equation of a relationship when you are given its graph. Use an example in your answer.

6.10 Two Linear Relations

Speed skating is an Olympic sport that has grown in popularity.

Investigate

Exploring a Pair of Relations

Mei and Zachary are junior speed skaters.
Mei skates at an average speed of 10 m/s.
Zachary skates at an average speed of 9 m/s.
Mei and Zachary have a 1500-m race.
Zachary has a head start of 100 m.
Who wins the race? How do you know?

Reflect

➤ How did you solve the problem?
If you did *not* draw a graph, work with a partner to use a graph to solve the problem.

➤ Suppose the race was only 500 m.
Who would have won? How do you know?

➤ Suppose the race was 1000 m.
Who would have won? How do you know?

Connect the Ideas

Recall that
1 km = 1000 m
So, 2.5 km = 2500 m

Diana and Kim live 2.5 km apart.
They leave their homes at the same time.
They travel toward each other's house.
Diana is on rollerblades.
She skates at an average speed of 250 m/min.
Kim is walking at an average speed of 60 m/min.
We can draw a graph to determine where and when they will meet.

The table below shows the distance each person is from Kim's house.
Each minute, Kim walks 60 m and Diana skates 250 m.
Graph the data for each person on the same grid.

Time (min)	Kim's distance (m)	Diana's distance (m)
0	0	2500
1	60	2250
2	120	2000
3	180	1750
4	240	1500

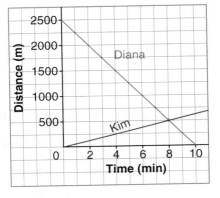

How else could this problem have been solved?

The lines intersect at approximately (8, 500).
The point of intersection shows where and when the people meet: after about 8 min and about 500 m from Kim's house.

Practice

1. One hot air balloon is descending.
   Another is ascending.
   The graph shows their heights over time.
   a) What is the starting height of the descending balloon?
   b) Where do the lines intersect?
      What does this point represent?

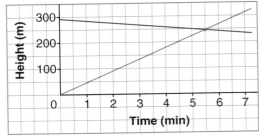

242   CHAPTER 6: Linear Relations

2. Nyla and Richard are travelling toward each other along the same bike path.
   Nyla jogs at an average speed of 100 m/min.
   Richard bikes at an average speed of 300 m/min.
   The table shows their distances, in metres,
   relative to Richard's starting position.

Time (min)	Nyla's distance (m)	Richard's distance (m)
0	4000	0
5		
10		

   a) To begin, how far apart are they?
   b) Copy and complete the table for times up to 30 min.
      Graph the data.
   c) Where do the lines intersect? What does this point represent?

3. **Assessment Focus** Hatef and Sophie are participating in a swim-a-thon.
   Swimmers may complete a maximum of 50 laps.
   Hatef has collected pledges totalling $13 per lap.
   Sophie has collected $96 in cash and pledges totalling $5 per lap.
   a) Make a table to show how much money each swimmer collects
      for completing up to 50 laps. Graph both sets of data on the same grid.
   b) What is the most money each person can collect?
   c) Where do the lines intersect? What does this point represent?

Two lines on a graph can be used to help make business decisions.

> **Example**
>
> Julie started a business making bracelets.
> Her costs are $180 for tools and advertising, plus $2 in materials
> per bracelet. Each bracelet is sold for $17.
> a) Write an equation for the total cost, C dollars, to make n bracelets.
> b) Write an equation for the revenue, R dollars,
>   from the sale of n bracelets.
> c) Graph both equations on the same grid.
> d) Where do the lines intersect?
>   What does this point represent?
>
> *Revenue is the money Julie gets when she sells the bracelets.*
>
> **Solution**
> a) Total cost = (initial costs) + (materials cost)
>   The initial costs are $180.
>   The materials cost is $2 per bracelet.
>   The equation is: $C = 180 + 2n$
> b) Revenue = (selling price per bracelet) × (number of bracelets sold)
>   The selling price is $17, so $R = 17n$
> c) To draw each graph, use the vertical intercept and one other point.
>   For *Total cost*, the vertical intercept is 180.
>   Plot the point (0, 180).
>   Substitute $n = 10$ in the equation $C = 180 + 2n$.
> $C = 180 + (2 \times 10) = 200$
>   Plot the point (10, 200).

6.10 Two Linear Relations

Join the points with a broken line and extend the line to the right.
For *Revenue*, the vertical intercept is 0.
Plot the point (0, 0).
Substitute $n = 10$ in the equation $R = 17n$.
$R = 17 \times 10 = 170$
Plot the point (10, 170).
Join the points with a broken line and extend the line to the right.

Julie's Jewellery Business

d) The lines intersect at approximately (12, 200).
This is the point where the total cost equals the revenue.
To the left of this point, the *Revenue* graph is below the *Total cost* graph, so Julie makes a loss.
To the right of this point, the *Revenue* graph is above the *Total cost* graph, so Julie makes a profit.

4. The weekly cost of running an ice-cream cart is a fixed cost of $150, plus $0.25 per ice-cream treat. The revenue is $2.25 per treat.
   a) Write an equation for the weekly cost, C dollars, when n treats are sold.
   b) Write an equation for the revenue, R dollars, when n treats are sold.
   c) Graph the equations on the same grid.
   d) How many treats have to be sold before a profit is made? Explain.

5. **Take It Further** Mark reads at an average rate of 30 pages per hour.
   Vanessa reads at an average rate of 40 pages per hour.
   Mark started reading a book.
   Vanessa started reading the same book when Mark was on page 25.
   a) How many hours later will they be reading the same page? How do you know?
   b) How many different ways can you answer part a? Describe each way.

In Your Own Words

Describe a situation that can be modelled by two lines on a graph.
Explain what the point of intersection represents.

Using a Graphing Calculator to Determine Where Two Lines Meet

You will need a TI-83 or TI-84 graphing calculator.

Nicole joins a fitness club.
The club offers two types of memberships.
Nicole can pay a monthly fee of $24, plus $2 for each visit.
Or, Nicole can pay $8 per visit.
Which type of membership should Nicole buy?

Let C dollars represent the total monthly cost.
Let n represent the number of times Nicole uses the gym in a month.
The equations are:

$C = 24 + 2n$ $\qquad\qquad\qquad C = 8n$

The graphing calculator uses X to represent n and Y to represent C.
To match this style, the equations are written as:

$Y = 24 + 2X$ $\qquad\qquad\qquad Y = 8X$

Follow these steps to graph the equations and determine the point of intersection.

To make sure there are no equations in the [Y=] list: Press [Y=]. Move the cursor to any equation, then press [CLEAR]. When all equations have been cleared, position the cursor to the right of the equals sign for Y1. To enter the equations, press: [2] [4] [+] [2] [X,T,Θ,n] [ENTER] [8] [X,T,Θ,n]	Plot1 Plot2 Plot3 \Y1■24+2X \Y2■8X \Y3= \Y4= \Y5= \Y6= \Y7=
To adjust the window settings to display the graph: Press [WINDOW]. Set the plot options as shown. To set an option, clear what is there, enter the value you want, then press [ENTER]. Use the ⊙ key to scroll down the list.	WINDOW Xmin=0 Xmax=6 Xscl=1 Ymin=0 Ymax=50 Yscl=5 Xres=1■

We are only interested in positive values of X and Y.

Technology: Using a Graphing Calculator to Determine Where Two Lines Meet

Press [GRAPH]. A graph like the one shown here should be displayed. The points lie on a line. Since X is a whole number, each line should be a series of dots joined with a broken line. However, the calculator joins the points with solid lines.	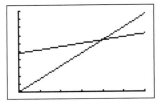
To determine the point of intersection: Press [2nd] [TRACE] to access the CALCULATE menu. Then press [5]. Your screen should look like the one shown here. Press ▶ to move the flashing point until it is close to the point of intersection. Press [ENTER] [ENTER] [ENTER].	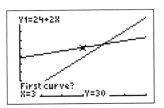
Your screen should look like the one shown here. The coordinates of the point of intersection appear at the bottom of the screen. • Where do the lines intersect? • What does this point represent? • Which type of membership would you recommend to Nicole? Explain your thinking.	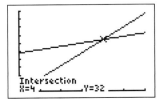

Use pages 245 and 246 as a guide.

Tyresse practises at the driving range to improve his golf swing. The driving range offers two payment options:
- Tyresse can pay $6.50 for each bucket of balls he hits.
- Tyresse can purchase a membership for $24.50. He will then pay $4.75 for each bucket of balls he hits.

➤ Write an equation to represent each payment option.
➤ Use a TI-83 or TI-84 calculator to graph the equations, using appropriate window settings.
➤ Determine where the lines intersect. Explain what the point represents.
➤ Which payment option would you recommend to Tyresse? Explain your thinking.

Chapter Review

What Do I Need to Know?

Both direct variation and partial variation are linear relations.

➤ Direct Variation

A graph that represents direct variation is a straight line that passes through the origin.

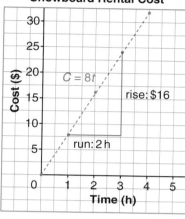

Time t (h)	Cost C ($)	First Differences
0	0	
1	8	$8 - 0 = 8$
2	16	$16 - 8 = 8$
3	24	$24 - 16 = 8$
4	32	$32 - 24 = 8$

Rate of change = $\frac{\text{rise}}{\text{run}}$

The equation is: $C = 8t$
↑
rate of change is $8/h

Use Frayer models to show what you know about direct variation and partial variation.

➤ Partial Variation

A graph that represents partial variation is a straight line that does not pass through the origin.

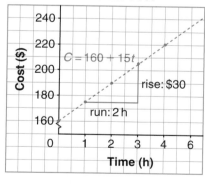

Time t (h)	Cost C ($)	First Differences
0	160	
1	175	$175 - 160 = 15$
2	190	$190 - 175 = 15$
3	205	$205 - 190 = 15$
4	220	$220 - 205 = 15$

Rate of change = $\frac{\text{rise}}{\text{run}}$

       fixed cost   variable cost
              ↘   ↘
The equation is: $C = 160 + 15t$
              ↗   ↖
vertical intercept is $160   rate of change is $15/h

Chapter Review **247**

What Should I Be Able to Do?

6.1 1. This pattern continues.

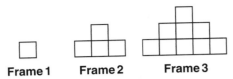

Frame 1   Frame 2   Frame 3

   a) Sketch the next frame in the pattern.
   b) Copy and complete the table below for the first 6 frames.

Frame number	Number of squares	First differences
1		
2		
3		

   c) Is the relationship linear or non-linear? Explain.
   d) Graph the relationship. Does the graph support your answer to part c? Explain.

6.2
6.3 2. Determine each rate of change. Explain what it means.

   a) Car Journey

   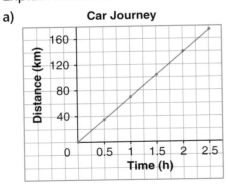

   b) How Water Evaporates

   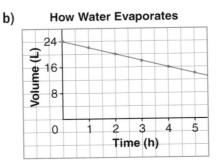

6.4
6.5 3. Does each graph in question 2 represent direct variation or partial variation? How do you know?

6.4 4. Ken delivers packets of flyers for local stores. His pay varies directly as the number of packets delivered.

Number of packets delivered	0	100	200	300
Pay ($)	0	25	50	75

   a) Graph *Pay* against *Number of packets delivered*.
   b) Determine the pay for 250 packets delivered.
   c) Suppose Ken wants to earn $100. How many packets must he deliver?
   d) About how many packets does Ken have to deliver to earn $140?

6.5 5. The cost to rent a snowboard for an 8-h day is $28. If the snowboard is kept longer, there is an additional fee per hour.

Number of extra hours	0	1	2	3	4
Rental cost ($)	28	31	34	37	40

   a) Graph *Rental cost* against *Number of extra hours*.
   b) What does the vertical intercept represent?
   c) What is the rate of change? What does it represent?
   d) Write an equation to determine the rental cost, C dollars, for t extra hours.

6.6   **6.** To set up a hot air balloon, the crew inflates the balloon with cold air. One fan can blow 450 m³ of air in 1 min.
     a) Make a table of values for the volume, V cubic metres, of air in the balloon when the fan runs for up to 6 min.
     b) Graph the data.
     c) Write an equation to determine the volume V after t minutes.
     d) Another fan can blow only half as much air per minute.
        i) How does the graph change? Draw the new graph on the grid in part b.
        ii) Write the new equation.

6.7   **7.** Solve each equation. Which tools could you use to help you?
     a) $4x + 5 = 9$
     b) $3 + 2x = -5$
     c) $17 = 5 + 3x$
     d) $18 = 3 - 5x$
     e) $7x - 6 = 15$
     f) $6x + 7 = -23$

6.8   **8.** A small pizza with tomato sauce and cheese costs $6.00. Each additional topping costs $0.75.
     a) Write a rule for the cost of the pizza when you know how many additional toppings it has.
     b) Write an equation for the cost, C dollars, of a small pizza when n toppings are ordered.
     c) You have $10.00 to spend. How many toppings can you order? Use two different methods to find out. Show your work.

6.9   **9.** The length of a person's foot is approximately 15% of his height.
     a) What is the approximate foot length of a person who is 180 cm tall?
     b) What is the approximate height of a person whose foot is 21 cm long?
     c) How can you use an equation to answer these questions?
     d) What other methods could you use to answer parts a and b?

6.10   **10.** Jo-Anne is choosing an Internet service provider.
- Speed Dot Company costs $2.40 for every hour of Internet use.
- Communications Plus costs $12 per month plus $0.90 for each hour of Internet use.

     a) Write an equation to model the cost for each billing system. Use C to represent the total cost, in dollars, and t to represent the time in hours.
     b) Make a table of values up to 10 h for each relationship. Graph both relationships on the same grid.
     c) What are the coordinates of the point of intersection? What do they represent?
     d) Jo-Anne will use the Internet about 7 h per month. Which company should she choose? Explain your choice.

Practice Test

Multiple Choice: Choose the correct answer for questions 1 and 2.

1. Which equation represents partial variation?
   A. $C = 20n$ B. $C = 20n^2$ C. $C = 20n + 5$ D. $C = -20n$

2. Which equation has the solution $x = 5$?
   A. $2x + 6 = 7$ B. $2x + 5 = 9$ C. $2x + 4 = 11$ D. $2x + 3 = 13$

Show your work for questions 3 to 6.

3. **Knowledge and Understanding**
   Andy's pay varies directly as the number of hours he works.
   He earns $24 for working a 3-h shift.
   a) Graph Andy's pay against time.
   b) Determine the rate of change.
   c) How much does Andy earn in 36 h? Explain.
   d) How long does it take Andy to earn $216? Explain.

4. **Communication** The cost to rent a banquet hall is shown in the graph.
   Write 5 different things you can determine from the graph.

   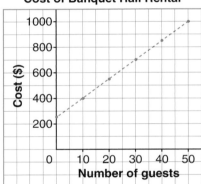

5. **Application**
   One ad in a local newspaper cost $70.
   The ad had 25 words.
   Another ad in the same newspaper cost $100.
   It had 40 words.
   How much would it cost to place an ad with 30 words?
   a) Use a graph to solve the problem.
   b) Use an equation to solve the problem.
   c) Which did you prefer: using a graph or using an equation? Explain.

6. **Thinking** The school store wants to distribute brochures describing its products.
   The store manager gets bids from two printers.
   Printer A charges an initial fee of $280 plus $1 per brochure.
   Printer B charges an initial fee of $90 plus $2 per brochure.
   Which printer should the manager choose? Explain.

7 Polynomials

What You'll Learn
To use different tools to simplify algebraic expressions and to solve equations

And Why
To provide further strategies for problem solving

Key Words
- zero pair
- term
- like terms
- polynomial
- constant term
- distributive law
- monomial
- coefficient

Project Link
- It's All in the Game

Reading and Writing in Math

Doing Your Best on a Test

Taking a math test may be stressful.

Here are some tips that can help you do your best without getting too stressed.

Getting Ready
- Find out what will be covered on the test.
- Give yourself lots of time to review.
- Look over your notes and examples. Make sure they are complete.
- Get help with difficult questions.
- Make study notes. Use a mind map or a Frayer model.

When You First Get the Test
- Listen to the instructions carefully.
- Ask questions about anything that is not clear.
- Look over the whole test before you start.
- Write any formulas you have memorized at the top of the page before you begin.

Answering Multiple-Choice Questions
- Read the question and the possible answers carefully.
- Eliminate any possible answer that cannot be correct.
- Answer the question, even if you have to guess.

Answering Other Questions
- Make three passes through the test:
    1. Answer all the quick and easy questions.
    2. Answer all the questions that you know how to do but take some work.
    3. Work through the more difficult questions.
- Read the questions carefully.
  Underline or highlight important information.
  Look for key words such as:
  compare, describe, determine, explain, justify
- Pace yourself! Don't get stuck in one place.

When You Have Finished
- Make sure you have answered all the questions.
- Look over your written solutions. Are they reasonable? Did you show all your thinking and give a complete solution?

252   CHAPTER 7: Polynomials

7.1 Like Terms and Unlike Terms

Algebra tiles are integer tiles and variable tiles.

Positive tiles
- ■ represents 1
- ▌ represents x
- ■ represents x^2

Negative tiles
- ▫ represents -1
- ▐ represents $-x$
- □ represents $-x^2$

Any two opposite tiles add to 0. They form a **zero pair**.

Investigate Using Algebra Tiles to Model Expressions

Work with a partner. Repeat each activity 5 times.

> Place some red algebra tiles on your desk.
> Group like tiles together.
> Describe the collection of tiles in words and as an algebraic expression.

Use at least 2 different kinds of red tiles. Vary the number and kind of tiles used.

> Place some red and blue tiles on your desk.
> Group like tiles together.
> Remove any zero pairs.
> Describe the remaining collection of tiles in words and as an algebraic expression.

Reflect

> What are like tiles?
> Why can they be grouped together?
> How would you represent like tiles algebraically?
> Why can you remove zero pairs?
> How do you know you have given the simplest name to a collection of tiles?

Connect the Ideas

To organize a collection of algebra tiles, we group like tiles.

There are two x^2-tiles, three x-tiles, and five 1-tiles.
These tiles represent the expression $2x^2 + 3x + 5$.

When a collection contains red and blue tiles, we group like tiles and remove zero pairs.

One $-x^2$-tile, three x-tiles, and two 1-tiles are left.
We write $-1x^2 + 3x + 2$.
We could also write $-x^2 + 3x + 2$.

The expression $-x^2 + 3x + 2$ has 3 **terms**: $-x^2$, $3x$, and 2

Terms are numbers, variables, or the products of numbers and variables.
Terms that are represented by like tiles are called **like terms**.

> The numerical part of a term is its **coefficient**. The coefficient of $3x^2$ is 3. It tells us that there are three x^2-tiles.

$-x^2$ and $3x^2$ are like terms.
Each term is modelled with x^2-tiles.
In each term, the variable x is raised to the exponent 2.

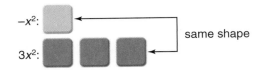

$-x^2$ and $3x$ are unlike terms.
Each term is modelled with different sized tiles.
Each term has the variable x, but the exponents are different.

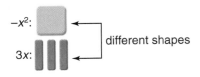

An expression is simplified when all like terms are combined, and any zero pairs are removed.

$-x^2 + 3x^2$ simplifies to $2x^2$.

$-x^2 + 3x$ cannot be simplified.

Practice

1. Which expression does each group of algebra tiles represent?
   a)
   b)
   c)
   d)
   e)
   f)

2. Use algebra tiles to model each expression. Sketch the tiles you used.
   a) $x - 5$
   b) $2x^2 + 3$
   c) $-x + 3$
   d) $x^2 - 4x$
   e) $4x^2 - 3x + 2$
   f) $-2x^2 - x - 5$

3. The diagram shows the length and width of the 1-tile, x-tile, and x^2-tile.

   Determine the area of each tile.
   Use your answer to explain the name of the tile.

4. Use algebra tiles to show $2x$ and $-4x$.
   Sketch the tiles you used.
   Are $2x$ and $-4x$ like terms? Explain.

5. Use algebra tiles to show $3x$ and $3x^2$.
   Sketch the tiles you used.
   Are $3x$ and $3x^2$ like terms? Explain.

6. a) Identify terms that are like $3x$:
      $-5x, 3x^2, 3, 4x, -11, 9x^2, -3x, 7x, x^3$
   b) Identify terms that are like $-2x^2$:
      $2x, -3x^2, 4, -2x, x^2, -2, 5, 3x^2$
   c) Explain how you identified like terms in parts a and b.

7. In each part, combine like terms. Write the simplified expression.
   a)
   b)
   c)
   d)
   e)
   f)

7.1 Like Terms and Unlike Terms

8. Combine like terms. Use algebra tiles.
   a) $3x + 1 + 2x + 3$ b) $3x^2 - 2x + 5x + 4x^2$ c) $2x^2 + 3x - 2x + 4 - x^2$

9. Write an expression with 5 terms that has only 2 terms when it is simplified.

When we need many tiles to simplify an expression, it is easier to use paper and pencil.

Example

Simplify.
$15x^2 - 2x + 5 + 10x - 8 - 9x^2$

Visualize algebra tiles.

Solution

$15x^2 - 2x + 5 + 10x - 8 - 9x^2$
$= 15x^2 - 9x^2 - 2x + 10x + 5 - 8$ Group like terms.
$= 6x^2 + 8x - 3$ To combine like terms, add their coefficients.

10. a) Simplify each expression.
    i) $-2 + 4x - 2x + 3$
    ii) $2x^2 - 3x + 4x^2 - 6x$
    iii) $3x^2 + 4x + 2 + x^2 + 2x + 1$
    iv) $x^2 - 4x + 3 - 2 + 5x - 4x^2$

    Which tools could you use to help you?

    b) Create an expression that cannot be simplified. Explain why it cannot be simplified.

11. **Assessment Focus**
    a) Determine the volume of this cube.
    b) Use the volume to suggest a name for the cube.
    c) Simplify. How can you use cubes to do this?
       i) $x^3 + 2x^3 + 5x^3$ ii) $3x^3 + 3x + x^3 + 5x$ iii) $5 - 2x^3 + 3x^2 + 5x^3$

12. **Take It Further** Many kits of algebra tiles contain a second variable tile called a y-tile.
    a) Why does a y-tile have a different length than an x-tile?
    b) Sketch algebra tiles to represent $2x - 5y - 1 + 4y - 7x + 4$.
    c) Write the expression in part b in simplest form. How do you know it is in simplest form?

In Your Own Words

Create a Frayer model for like terms.
Explain how like terms can be used to write an expression in simplest form.
Use diagrams and examples in your explanation.

7.2 Modelling the Sum of Two Polynomials

Expressions such as $3x^2$, $2x + 1$, and $4x^2 - 2x + 3$ are polynomials.
A **polynomial** can be one term, or the sum or difference of two or more terms.

Investigate — Using Algebra Tiles to Add Polynomials

Work with a partner.
You will need algebra tiles and a paper bag.

➤ Put the red algebra tiles in a bag.
Take turns to remove a handful of tiles,
then write the polynomial they model.
Add the two polynomials.
Write the resulting polynomial.

➤ Repeat the previous activity with red and blue tiles.

Reflect

➤ How do you model addition with algebra tiles?

➤ Which terms can be combined when you add polynomials?
Why can these terms be combined?

➤ Compare the original polynomials and the polynomial for their sum.
What remains the same? What changes? Explain.

7.2 Modelling the Sum of Two Polynomials **257**

Connect the Ideas

To add the polynomials $2x^2 - 5x + 3$ and $x^2 + 3x - 2$, we write $(2x^2 - 5x + 3) + (x^2 + 3x - 2)$.

Use algebra tiles

Model each polynomial.

$2x^2 - 5x + 3$

$x^2 + 3x - 2$

Combine like terms.

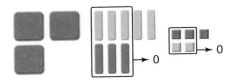

The remaining tiles represent $3x^2 - 2x + 1$.
So, $(2x^2 - 5x + 3) + (x^2 + 3x - 2) = 3x^2 - 2x + 1$

Use paper and pencil

Group like terms and simplify.

$(2x^2 - 5x + 3) + (x^2 + 3x - 2)$ Remove the brackets.
$= 2x^2 - 5x + 3 + x^2 + 3x - 2$ Group like terms.
$= 2x^2 + x^2 - 5x + 3x + 3 - 2$ Combine like terms.
$= 3x^2 - 2x + 1$

You can also add polynomials vertically.
$(2x^2 - 5x + 3) + (x^2 + 3x - 2)$

Align like terms.

$$\begin{array}{r} 2x^2 - 5x + 3 \\ +\ x^2 + 3x - 2 \\ \hline 3x^2 - 2x + 1 \end{array}$$

Combine like terms.

Practice

1. Which polynomial sum does each set of tiles represent?
   a)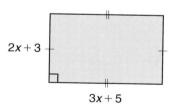
   b)

2. Use algebra tiles to add these polynomials. Sketch the tiles you used.
   a) $(2 + x^2) + (-3x^2 - 5)$
   b) $(-2 - x^2) + (3x^2 + 5)$
   c) $(-2 + x^2) + (-3x^2 + 5)$
   d) $(2 - x^2) + (-3x^2 + 5)$

3. Add. Use algebra tiles if it helps.
   a) $(2x - 3) + (4x + 5)$
   b) $(9x - 5) + (7 - 6x)$
   c) $(-x + 2) + (5x - 1)$
   d) $(3x + 3) + (-4x - 5)$
   e) $(8x - 2) + (-6x - 6)$
   f) $(3x - 2) + (-x + 6)$

4. Add. Use algebra tiles if it helps.
   a) $(3x^2 - 4x) + (5x^2 + 7x - 1)$
   b) $(-3x^2 - 4x + 2) + (5x^2 - 8x - 7)$
   c) $(x^2 - 2x - 1) + (3x^2 + x + 2)$
   d) $(x^2 - 5) + (-2x^2 + 2x - 3)$
   e) $(3x^2 - x + 5) + (-8x^2 + 3x - 1)$
   f) $(3x^2 + 2x + 1) + (-2x^2 + 3x - 4)$

5. Add: $(3x^3 + 2x^2 - x + 1) + (-2x^3 - 6x^2 + 4x - 3)$
   Explain how you did it.

   Which tools could you use to help you?

Polynomials can be used to represent side lengths of figures.

Example

Write a simplified expression for the perimeter of this rectangle.

Solution The perimeter is the sum of the measures of the four sides.

$$2x + 3$$
$$+\ 2x + 3$$
$$+\ 3x + 5$$
$$+\ 3x + 5$$
$$\overline{\ 10x + 16\ }$$

The perimeter is $10x + 16$.

6. Write an expression for the perimeter of each figure.
   Simplify the expression.

   a)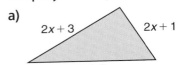

   b)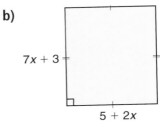

   c)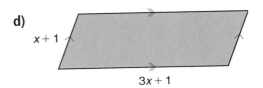

   d)

   Need Help?
   Read the Example on page 259.

7. Choose one of the figures in question 6.
   a) Evaluate the unsimplified expression for the perimeter when $x = 8$.
   b) Evaluate the simplified expression for the perimeter when $x = 8$.
   c) Is it better to substitute into an expression before it is simplified or after it is simplified? Explain.

8. Create a polynomial that is added to $2x^2 + 3x + 7$ to get each sum.
   a) $4x^2 + 5x + 9$
   b) $3x^2 + 4x + 8$
   c) $2x^2 + 3x + 7$
   d) $x^2 + 2x + 6$
   e) $x + 5$
   f) 4

9. **Assessment Focus** Two polynomials are added.
   Their sum is $3x^2 - 2x + 5$.
   Write two polynomials that have this sum.
   How many different pairs of polynomials can you find?
   Which tools did you use to help you?

10. **Take It Further** The sum of 2 polynomials is $2x^2 - 7x + 3$.
    One polynomial is $3x^2 - 5x - 2$.
    What is the other polynomial?
    Explain how you found it.

In Your Own Words

Write two things you know about adding polynomials.
Use an example to illustrate.

Using CAS to Add and Subtract Polynomials

You will need a TI-89 calculator.

Step 1

- Clear the home screen:
  HOME F1 8 CLEAR

- Enter $1 + 1 + 1$:
  1 + 1 + 1 ENTER

- Enter $x + x + x$:
  X + X + X ENTER

- Enter $x^2 + x^2 + x^2$:
  X ^ 2 + X ^ 2 + X ^ 2 ENTER

- Enter $1 + x + x^2$:
  1 + X + X ^ 2 ENTER

At the end of Step 1, your screen should look like this:

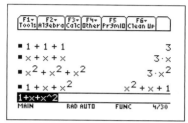

Record the first 3 lines of the screen on paper. What patterns do you see?

Use algebra tiles to explain the patterns.

Record the last line of the screen on paper. Use algebra tiles to explain why the terms could not be simplified.

Step 2

To add $6x + 2$ and $3x + 4$, we write: $(6x + 2) + (3x + 4)$

- Clear the home screen:
  HOME F1 8 CLEAR

- Enter $(6x + 2) + (3x + 4)$:
  (6X + 2) + (3X + 4) ENTER

- Enter $(3x + 5) + (2x + 3)$:
  (3X + 5) + (2X + 3) ENTER

- Enter $(2x + 7) + (5x - 3)$.

- Enter $(4x - 8) + (-7x + 2)$.

At the end of Step 2, your screen should look like this:

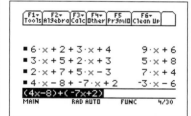

Record the screen on paper.

Which terms are combined when polynomials are added?

What remains the same when the terms are combined?

What changes when the terms are combined?

Step 3

Use paper and pencil to add these polynomials:

- $(5x + 6) + (8x + 3)$
- $(3x - 3) + (2x + 5)$

Use CAS to check your work.

Explain any mistakes you made. How could you avoid making the same mistakes in the future?

Step 4

- Clear the home screen:
  [HOME] [F1] 8 [CLEAR]

- Enter $3x - x$:
  3[X] [−] [X] [ENTER]

- Enter $3x - 2x$:
  3[X] [−] 2[X] [ENTER]

- Enter $3x - 3x$:
  3[X] [−] 3[X] [ENTER]

- Continue the pattern until your screen is like the one at the right.

At the end of Step 4, your screen should look like this:

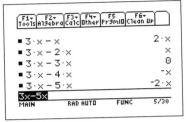

Record the screen on paper.

What remains the same when terms are subtracted?

What changes when terms are subtracted?

Step 5

To subtract $3x + 2$ from $6x + 4$, we write:
$(6x + 4) - (3x + 2)$

- Clear the home screen:
  [HOME] [F1] 8 [CLEAR]

- Enter $(6x + 4) - (3x + 2)$:
  [(]6[X] [+] 4[)] [−] [(]3[X] [+] 2[)] [ENTER]

- Enter $(3x + 5) - (2x + 3)$:
  [(]3[X] [+] 5[)] [−] [(]2[X] [+] 3[)] [ENTER]

- Enter $(2x + 3) - (5x + 7)$.

- Enter $(4x - 8) - (-x + 2)$.

At the end of Step 5, your screen should look like this:

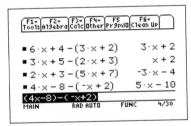

Record the screen on paper.

Explain each result.

Step 6

Use CAS to subtract these polynomials:

- $(x - 8) - (2x + 3)$
- $(-3x + 5) - (-2x - 1)$

Explain any mistakes you made. How could you avoid making the same mistakes in the future?

7.3 Modelling the Difference of Two Polynomials

Investigate — Using Algebra Tiles to Subtract Polynomials

Work with a partner.
You will need algebra tiles.

➤ Subtract.
  $6 - 4$
  $5x - 2x$
  $(4x + 6) - (2x + 3)$
  $(2x^2 + 5x + 3) - (x^2 + 2x + 2)$

➤ Subtract.
  Add zero pairs so you have enough tiles to subtract.
  $4 - 8$
  $2x - 3x$
  $(2x + 4) - (3x + 8)$
  $(x^2 + 2x + 1) - (3x^2 + 4x + 3)$

Reflect

➤ How did you use algebra tiles to subtract polynomials?

➤ Why can you use zero pairs to subtract?

➤ How do you know how many zero pairs you need, to be able to subtract?

7.3 Modelling the Difference of Two Polynomials **263**

Connect the Ideas

To determine: $(-x^2 - 2x + 1) - (2x^2 + 4x - 3)$

Use algebra tiles to model $-x^2 - 2x + 1$:

To subtract $2x^2 + 4x - 3$, we need to take away 2 red x^2-tiles, 4 red x-tiles, and 3 blue 1-tiles. But we do not have these tiles to take away. So, we add zero pairs that include these tiles.

Zero pair for $2x^2$:

Zero pair for $4x$:

Zero pair for -3:

> Adding these tiles has not changed the polynomial represented.

Now we can take away the tiles for $2x^2 + 4x - 3$.

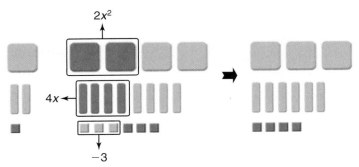

The remaining tiles represent $-3x^2 - 6x + 4$.
So, $(-x^2 - 2x + 1) - (2x^2 + 4x - 3) = -3x^2 - 6x + 4$

When we subtracted $2x^2 + 4x - 3$, we ended up adding a polynomial with opposite signs. That is, we added the opposite polynomial, $-2x^2 - 4x + 3$:

> We get the opposite polynomial by changing all the signs of the coefficients of a polynomial.

This is true in general.
To subtract a polynomial, we add its opposite.

Practice

1. Which polynomial difference is represented by these tiles?

2. Simplify these polynomials. Use algebra tiles.
   a) $(5x + 3) - (3x + 2)$
   b) $(5x + 3) - (3x - 2)$
   c) $(5x + 3) - (-3x + 2)$
   d) $(5x + 3) - (-3x - 2)$

3. Simplify these polynomials. Use algebra tiles.
   a) $(3x^2 + 2x + 4) - (2x^2 + x + 1)$
   b) $(3x^2 - 2x + 4) - (2x^2 - x + 1)$
   c) $(3x^2 - 2x - 4) - (-2x^2 + x - 1)$
   d) $(-3x^2 + 2x - 4) - (2x^2 - x - 1)$

4. Show the opposite of each polynomial using algebra tiles. Write the opposite polynomial.
   a) $3x - 2$
   b) $-2x^2 + 3x$
   c) $4x^2 - 7x + 6$
   d) $-x^2 + 5x - 4$

We can subtract polynomials using paper and pencil.

Example

Subtract: $(5x^2 - 2x + 4) - (7x^2 + 10x - 8)$

Solution

$(5x^2 - 2x + 4) - (7x^2 + 10x - 8)$

To subtract $7x^2 + 10x - 8$, add its opposite $-7x^2 - 10x + 8$.

$(5x^2 - 2x + 4) - (7x^2 + 10x - 8)$
$= 5x^2 - 2x + 4 + (-7x^2 - 10x + 8)$
$= 5x^2 - 2x + 4 - 7x^2 - 10x + 8$
$= 5x^2 - 7x^2 - 2x - 10x + 4 + 8$
$= -2x^2 - 12x + 12$

To check, add the difference to the second polynomial:
$(-2x^2 - 12x + 12) + (7x^2 + 10x - 8)$
$= -2x^2 + 7x^2 - 12x + 10x + 12 - 8$
$= 5x^2 - 2x + 4$

The sum is equal to the first polynomial.
So, the difference is correct.

5. Simplify.
   a) $(x + 7) - (x + 5)$
   b) $(x + 7) - (x - 5)$
   c) $(x + 7) - (-x + 5)$
   d) $(x + 7) - (-x - 5)$

6. Simplify. Check your answers by adding.
   a) $(2x^2 - 3) - (x^2 + 1)$
   b) $(3x^2 + 2x) - (2x^2 + x)$
   c) $(7 - 4x^2) - (8 - 2x^2)$
   d) $(5x - 7x^2) - (2x^2 + 2x)$

   Which tools can you use to help you?

7. Simplify. How could you check your answers?
   a) $(3x^2 + x - 1) - (x^2 - 2x + 5)$
   b) $(x^2 - x + 1) - (x^2 + x - 1)$
   c) $(2x^2 + x - 3) - (x^2 - 3x + 4)$
   d) $(x - x^3 + 5) - (7 - x + x^3)$
   e) $(7 + 3x - 2x^3) - (4 - 3x + 3x^3)$
   f) $(2x^2 - 3x - 5) - (2x^2 - 3x - 5)$

8. **Assessment Focus** John subtracted these polynomials:
   $(2x^2 - 4x + 6) - (3x^2 + 2x - 4)$
   a) Explain why his solution is incorrect.
   $$(2x^2 - 4x + 6) - (3x^2 + 2x - 4)$$
   $$= 2x^2 - 4x + 6 - 3x^2 + 2x - 4$$
   $$= 2x^2 - 3x^2 - 4x + 2x + 6 - 4$$
   $$= -x^2 - 2x + 2$$
   b) What is the correct answer? Show your work.
   c) How could you check your answer?

9. a) Simplify.
      i) $(3x^2 + x) - (2x^2 - 3x)$
      ii) $(2x^2 - 3x) - (3x^2 + x)$
      iii) $(4x^3 - 5) - (7 + 2x^3)$
      iv) $(7 + 2x^3) - (4x^3 - 5)$
      v) $(2x - x^3) - (-x^3 + 2x)$
      vi) $(-x^3 + 2x) - (2x - x^3)$
   b) What patterns do you see in the answers in part a? Explain.
   c) Write two polynomials.
      Subtract them in different orders.
      What do you notice?

10. a) i) Write a polynomial.
       ii) Write the opposite polynomial.
    b) Subtract the two polynomials in part a.
       What do you notice about your answer?
    c) Compare your answer with those of your classmates.
       Is there a pattern? Explain.

11. **Take It Further** One polynomial is subtracted from another.
    The difference is $-2x^2 + 4x - 5$.
    Write two polynomials that have this difference.
    How many different pairs of polynomials can you find?

In Your Own Words

What did you find difficult about subtracting two polynomials?
Use examples to show how you overcame this difficulty.

I Have... Who Has?

A game card looks like this:

> I have ...
> $x + 1$
> Who has ...
> $5x - 2 + 3x - 2$ simplified?

Materials
2 sets of game cards: green for Team A and yellow for Team B

➤ Divide the class into two teams.
   Toss a coin to see which team starts.
   Distribute a game card to each student.

➤ A student from the team that won the coin toss starts.
   The student reads her card.
   "I have ... $x + 1$.
   Who has ... $5x - 2 + 3x - 2$ simplified?"

➤ All students on the second team simplify the expression.
   The student whose card has the solution $8x - 4$ says:
   "I have ... $8x - 4$.
   Who has ... the opposite of $3x^2 - 4x + 1$?"

➤ A team makes a mistake when a student does not respond or responds incorrectly.

➤ The game ends when all the cards have been called.
   The team with the fewer mistakes wins.

Mid-Chapter Review

7.1 1. Renata simplified an expression and got $-7 + 4x^2 + 2x$.
Her friend simplified the same expression and got $4x^2 + 2x - 7$.
Renata thinks her answer is wrong. Do you agree? Explain.

2. Which expression does each group of tiles represent?
   a)
   b)
   c)
   d)

3. Combine like terms.
   a) $5x + 6 + 3x$
   b) $3x - x + 2$
   c) $2x^2 - 6 + x^2 + 3$
   d) $-5 + x + 3 - 2x$
   e) $x - 4 + 6 - 5x$
   f) $3x^3 - x + 2x^3 + 3x$
   g) $4x^2 + 3x + 5 - 5x^2 - 2x - 7$

   Which tools could you use to help you?

4. Cooper thinks $5x + 2$ simplifies to $7x$. Is he correct? Explain.
   Sketch algebra tiles to support your explanation.

5. Write an expression with 4 terms that has only 1 term when it is simplified.
   Explain your thinking.

7.2 6. Which polynomial sum do these tiles represent?

7. Add.
   a) $(5x + 4) + (3x + 5)$
   b) $(4x - 3) + (5 - 2x)$
   c) $(7x^2 - 4x) + (3x^2 - 2)$
   d) $(5x^2 + 2x + 1) + (x^2 - 3x - 5)$
   e) $(-2x^2 - x + 1) + (x^2 + 4x - 2)$
   f) $(3x^2 + 4x + 2) + (-4x^2 - 3x - 5)$

8. Which polynomial must be added to $5x^2 + 3x - 2$ to get $7x^2 + 5x + 1$?

9. Write a simplified expression for the perimeter of each figure.
   a)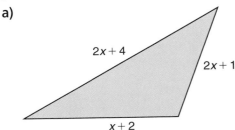
   b)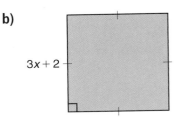

7.3 10. Write the opposite polynomial.
   a) $x + 3$
   b) $3x^2 - 5x$
   c) $2x^2 + 3x + 7$
   d) $-5x^2 - 2x - 1$

11. Subtract.
    How could you check your answers?
    a) $(x + 3) - (x - 4)$
    b) $(7x^2 - 4x) - (x^2 - 2x)$
    c) $(3x^2 - 2x + 7) - (5x^2 - 3x - 2)$
    d) $(-2x^2 - 4x + 1) - (3x^2 + 2x - 1)$
    e) $(4x^2 - 2x - 1) - (4x^2 - 5x - 8)$

7.4 Modelling the Product of a Polynomial and a Constant Term

Algebra tiles are area tiles.
The name of the tile tells its area.

Tile	Dimensions (units)	Area (square units)
1-tile	1 by 1	$(1)(1) = 1$
x-tile	1 by x	$(1)(x) = x$
x^2-tile	x by x	$(x)(x) = x^2$

Investigate — Using Algebra Tiles to Multiply

Work with a partner.
Use algebra tiles.
In each case, sketch the tiles you used.

Remember that multiplication is repeated addition.

➢ Display the polynomial $3x + 1$.
 Use more tiles to display the product $2(3x + 1)$.
 Combine like tiles.
 Write the resulting product as a polynomial.

➢ Display the product $3(3x + 1)$.
 Combine like tiles.
 Write the resulting product as a polynomial.

➢ Follow the steps above for these products:
 $4(3x + 1)$
 $5(3x + 1)$

Reflect

➢ What patterns do you see in the polynomials and their products?

➢ Use these patterns to write the product $6(3x + 1)$ as a polynomial.

➢ Use the patterns to find a way to multiply without using algebra tiles.

Connect the Ideas

Use a model

We can model $3(2x + 4)$ as the area of a rectangle in two ways.
The rectangle has length $2x + 4$ and width 3.

- Use algebra tiles to label the length and width.
- Use the labels as a guide. Fill in the rectangle.

Six x-tiles and twelve 1-tiles have a combined area of $6x + 12$.
So, $3(2x + 4) = 6x + 12$

- Draw and label a rectangle.
  The length of the rectangle is $2x + 4$.
  The width of the rectangle is 3.
  So, the area of the rectangle is $3(2x + 4)$.

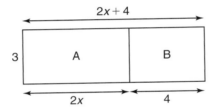

We divide the rectangle into two rectangles.
Rectangle A has area: $3(2x) = 6x$
Rectangle B has area: $3(4) = 12$
The total area is: $6x + 12$
So, $3(2x + 4) = 6x + 12$

Use paper and pencil

We can also determine $3(2x + 4)$ using paper and pencil.

Each term in the brackets is multiplied by the constant term 3 outside the brackets.
This process is called *expanding*.

$$3(2x + 4) = 3(2x) + 3(4)$$
$$= 6x + 12$$

In algebra, a term without a variable is called a **constant term**.

This illustrates the **distributive law**.
When we use the distributive law to multiply a polynomial by a constant term, we *expand* the product.

Practice

1. Write the product modelled by each set of tiles. Determine the product.
   a)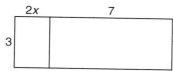
   b)
   c)

2. Use algebra tiles to determine each product. Sketch the tiles you used.
   a) $2(4x + 1)$
   b) $2(3x + 1)$
   c) $5(2x + 3)$
   d) $4(4x + 3)$

3. Write the product modelled by each figure. Determine the product.
   a)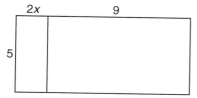
   b)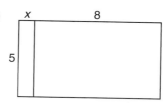
   c)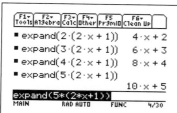
   d)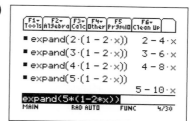

4. Use the calculator screens below.
   What patterns do you see? Explain each pattern.
   Write the next 2 lines in each pattern.
   a)
   ```
   ■ expand(2·(2·x + 1))    4·x + 2
   ■ expand(3·(2·x + 1))    6·x + 3
   ■ expand(4·(2·x + 1))    8·x + 4
   ■ expand(5·(2·x + 1))
                           10·x + 5
   expand(5*(2*x+1))
   ```
   b)
   ```
   ■ expand(2·(1 − 2·x))    2 − 4·x
   ■ expand(3·(1 − 2·x))    3 − 6·x
   ■ expand(4·(1 − 2·x))    4 − 8·x
   ■ expand(5·(1 − 2·x))
                           5 − 10·x
   expand(5*(1−2*x))
   ```

We can also use paper and pencil to expand a product.

> **Example**
> Expand: $-3(-2x^2 + 3x - 4)$
>
> **Solution** $-3(-2x^2 + 3x - 4)$
> Multiply each term in the brackets by -3.
> $-3(-2x^2 + 3x - 4) = (-3)(-2x^2) + (-3)(3x) + (-3)(-4)$
> $= (+6x^2) + (-9x) + (+12)$
> $= 6x^2 - 9x + 12$

5. Expand: $-100(4x^2 + x - 4)$
   Visualize algebra tiles. Are they useful to determine this product? Explain.

6. Jessica expands $3(x^2 + 4x - 2)$ and gets $3x^2 + 4x - 2$.
   Choose a tool.
   Use the tool to explain why Jessica's answer is incorrect.

7. Multiply.
   a) $7(3x - 1)$
   b) $3(4x - 5)$
   c) $2(6x - 4)$
   d) $5(5x^2 - 3x)$
   e) $4(-2x^2 + 3)$
   f) $9(x^2 + x - 6)$

8. Expand.
   a) $-2(4x + 2)$
   b) $3(-5x^2 + 3)$
   c) $2(-x - 4)$
   d) $-7(x^2 - 5)$
   e) $-2(-5x^2 + x)$
   f) $6(3 - 2x)$

9. **Assessment Focus** Multiply. Which tools did you use? Explain.
   a) $-2(x^2 - 2x + 4)$
   b) $-3(-x^2 + 3x - 7)$
   c) $2(4x^3 - 2x^2 - x)$
   d) $8(3x^3 + 2x^2 - 3)$
   e) $-6(2x^2 - x + 5)$
   f) $4(x^2 - 3x - 3)$

10. Square A has side length $4x + 1$.
    Square B has side length that is 3 times as great as that of Square A.

    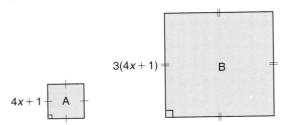

    a) Write an expression for the perimeter of each square.
       Simplify each expression.
    b) What is the difference in perimeters?

11. **Take It Further** Explain how you could use algebra tiles to multiply a polynomial by a negative constant term.
    Illustrate with an example.

In Your Own Words

When can you use algebra tiles to determine the product of a constant term and a polynomial?
When do you use paper and pencil?
Include examples in your explanation.

Using CAS to Investigate Multiplying Two Monomials

You will need a TI-89 calculator.

Step 1

- Clear the home screen:
  [HOME] [F1] 8 [CLEAR]
- Enter $x + x$:
  [X] [+] [X] [ENTER]
- Enter $x + x + x$:
  [▷] [+] [X] [ENTER]
- Continue the pattern of adding x until your screen is like the one at the right.

At the end of Step 1, your screen should look like this:

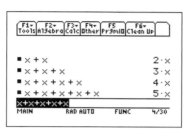

Record the screen on paper. Write the next 2 lines in the pattern.

Step 2

- Clear the home screen:
  [HOME] [F1] 8 [CLEAR]
- Enter x times x:
  [X] [×] [X] [ENTER]
- Enter x times x times x:
  [▷] [×] [X] [ENTER]
- Continue the pattern of multiplying by x until your screen is like the one at the right.

At the end of Step 2, your screen should look like this:

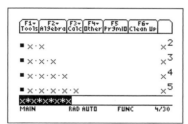

Record the screen on paper. Write the next 2 lines in the pattern.

Compare the results of Step 1 and Step 2. How are they different? Explain.

Step 3

- Clear the home screen:
  [HOME] [F1] 8 [CLEAR]
- Enter $x + 2x$:
  [X] [+] 2[X] [ENTER]
- Enter $2x + 3x$:
  2[X] [+] 3[X] [ENTER]
- Continue the pattern until your screen is like the one at the right.

At the end of Step 3, your screen should look like this:

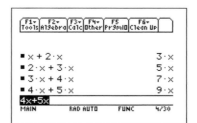

Record the screen on paper. Write the next 2 lines in the pattern.

Step 4

- Clear the home screen:
  [HOME] [F1] 8 [CLEAR]
- Multiply x by $2x$:
  [X] [×] 2[X] [ENTER]
- Multiply $2x$ by $3x$:
  2[X] [×] 3[X] [ENTER]
- Continue the pattern until your screen is like the one at the right.

At the end of Step 4, your screen should look like this:

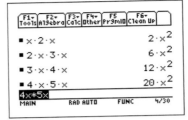

Record the screen on paper. Write the next 2 lines in the pattern.

A polynomial, such as $2x$ or $3x$, is called a monomial. How is multiplying monomials different from adding monomials?

Step 5

- Clear the home screen:
  [HOME] [F1] 8 [CLEAR]
- Multiply x by $2x^2$:
  [X] [×] 2[X][^]2 [ENTER]
- Multiply $2x$ by $3x^2$:
  2[X] [×] 3[X][^]2 [ENTER]
- Continue the pattern until your screen is like the one at the right.

At the end of Step 5, your screen should look like this:

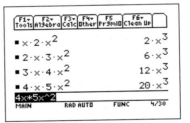

Record the screen on paper. Write the next 2 lines in the pattern. Explain the pattern.

Step 6

Use CAS to multiply.

- $(2x)(6x)$
- $(2x)(-6x)$
- $(-2x)(6x)$
- $(-2x)(-6x)$
- $(5x)(5x)$
- $(-8x)(3x^2)$
- $(-2x^2)(-3x)$

$(2x)(6x)$ means $2x$ multiplied by $6x$. How could you use paper and pencil to multiply two monomials?

7.5 Modelling the Product of Two Monomials

Investigate — Using Algebra Tiles to Multiply Monomials

Use algebra tiles.
In each case, sketch the tiles you used.

Recall that $(x)(2x)$ means x times $2x$.

- Build a rectangle x units wide and $2x$ units long.
  Which tiles did you use?
  What is the area of the rectangle?
  Simplify: $(x)(2x)$

- Show the product: $(2x)(2x)$
  Simplify: $(2x)(2x)$

- Show the product: $(3x)(2x)$
  Simplify: $(3x)(2x)$

- Show the product: $(4x)(2x)$
  Simplify: $(4x)(2x)$

- Show the product: $(5x)(2x)$
  Simplify: $(5x)(2x)$

Reflect

- What patterns do you see in the terms and their products?
- Use these patterns to simplify: $(6x)(2x)$
- Use the patterns to find a way to multiply without using algebra tiles.

7.5 Modelling the Product of Two Monomials **275**

Connect the Ideas

A polynomial with one term is called a **monomial**.
Here are some monomials: $2x, 5, -3, 6x^2, -x^3$

Use algebra tiles

We can use algebra tiles to determine the product of some monomials.

To determine $(3x)(4x)$, draw two sides of a rectangle whose length is $4x$ and whose width is $3x$.

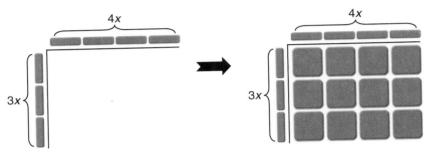

Use algebra tiles to fill the rectangle.
We need twelve x^2-tiles.
So, $(3x)(4x) = 12x^2$

Look at the product:
$(3x)(4x)$

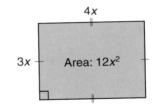

The numbers 3 and 4 are coefficients.
To multiply two monomials,
    multiply the coefficients: $(3)(4) = 12$
    multiply the variables: $(x)(x) = x^2$
So, $(3x)(4x) = 12x^2$

Use paper and pencil

We cannot use algebra tiles to determine a product such as:
$(x)(2x^2)$
We use paper and pencil instead.
Multiply the coefficients: $(1)(2) = 2$
Multiply the variables: $(x)(x^2) = (x)(x)(x)$
                                                      $= x^3$

So, $(x)(2x^2) = 2x^3$

The term x has coefficient 1; that is, x can be written as $1x$.

Practice

1. Write the product modelled by each set of tiles.
Determine the product.

a)   b)  c)

2. Look at the statements and products in question 1.
How are they alike?
How are they different?

3. Write the product modelled by each set of tiles.
Determine the product.

a)   b)   c)

Having Trouble?
Read Connect the Ideas

4. Use algebra tiles to multiply.
a) $(x)(3x)$ b) $(x)(5x)$ c) $(3x)(5x)$ d) $(4x)(x)$
e) $(x)(4x)$ f) $(2x)(6x)$ g) $(7x)(2x)$ h) $(3x)(3x)$

5. Look at the calculator screens below.
What patterns do you see?
Explain each pattern.
Write the next 3 lines in each pattern.

a)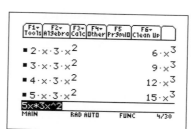

Screen a):
- $2 \cdot x \cdot 2 \cdot x \qquad 4 \cdot x^2$
- $2 \cdot x \cdot 3 \cdot x \qquad 6 \cdot x^2$
- $2 \cdot x \cdot 4 \cdot x \qquad 8 \cdot x^2$
- $2 \cdot x \cdot 5 \cdot x \qquad 10 \cdot x^2$

2x*5x

Screen b):
- $2 \cdot x \cdot 3 \cdot x^2 \qquad 6 \cdot x^3$
- $3 \cdot x \cdot 3 \cdot x^2 \qquad 9 \cdot x^3$
- $4 \cdot x \cdot 3 \cdot x^2 \qquad 12 \cdot x^3$
- $5 \cdot x \cdot 3 \cdot x^2 \qquad 15 \cdot x^3$

5x*3x^2

6. Multiply.
a) $(2x^2)(3x)$ b) $(4x)(x^2)$ c) $(x)(3x^2)$ d) $(x^2)(x)$
e) $(6x^2)(4x)$ f) $(2x)(8x^2)$ g) $(3x^2)(8x)$ h) $(3x)(3x^2)$

We can use paper and pencil to multiply monomials when there is a negative coefficient.

> **Example**
>
> Multiply: $(-3x^2)(7x)$
>
> **Solution** $(-3x^2)(7x)$
> Multiply the coefficients: $(-3)(7) = -21$
> Multiply the variables: $(x^2)(x) = (x)(x)(x)$
> $\qquad\qquad\qquad\qquad\qquad\quad = x^3$
>
> So, $(-3x^2)(7x) = -21x^3$

7. Multiply.
   a) $(-3)(2x)$ b) $(4x)(-4)$ c) $(8)(-x)$ d) $(5x)(-7)$
   e) $(2x)(-9)$ f) $(5)(-2x)$ g) $(-9x)(9)$ h) $(-3x)(-6)$

   Which tools can you use to help you?

8. Multiply.
   a) $(2x^2)(-3x)$ b) $(-3x)(4x^2)$ c) $(-4x)(-5x^2)$ d) $(x^2)(-x)$
   e) $(4x^2)(-8x)$ f) $(2x)(-2x^2)$ g) $(-x)(12x^2)$ h) $(-3x)(-12x^2)$

9. Students who have trouble with algebra often make these mistakes.
   ➤ They think $x + x$ equals x^2.
   ➤ They think $(x)(x)$ equals $2x$.
   Choose a tool.
   Use the tool to show how to get the correct answers.

10. **Assessment Focus**
    a) Determine each product.
       i) $(3x)(5x)$ ii) $(-3x)(5x)$ iii) $(3)(5x)$ iv) $(3x)(5)$
       v) $(5x^2)(3)$ vi) $(5x^2)(-3)$ vii) $(-5x^2)(-3)$ viii) $(5x^2)(3x)$
    b) List the products in part a that can be modelled with algebra tiles.
       Justify your list.
    c) Sketch the tiles for one product you listed in part b.

11. **Take It Further** The product of two monomials is $36x^3$.
    What might the two monomials be?
    How many different pairs of monomials can you find?

In Your Own Words

Suppose you have to multiply two monomials.
How do you know when you can use algebra tiles?
When do you have to use paper and pencil? When can you use CAS?
Include examples in your answer.

CHAPTER 7: Polynomials

7.6 Modelling the Product of a Polynomial and a Monomial

Investigate Using Models to Multiply a Polynomial by a Monomial

Work with a partner.

➤ Determine each product by using algebra tiles and by using an area model.
$2(x + 3)$
$2x(x + 3)$

➤ Write a different product that involves a monomial and a polynomial.
Have your partner model your product with algebra tiles and with a rectangle.

➤ Repeat the previous step for 4 other products.
Take turns to write the product, then model it.

Reflect

➤ What patterns do you see in the products you created and calculated?

➤ Did you write any products that could *not* be modelled with algebra tiles?
If your answer is yes, how did you determine the product?
If your answer is no, try to create a product that you cannot model with algebra tiles.
Try to find a way to determine the product.

Connect the Ideas

There are 3 ways to expand $3x(2x + 4)$.

Use algebra tiles

Make a rectangle with width $3x$ and length $2x + 4$.
Fill in the rectangle with tiles.

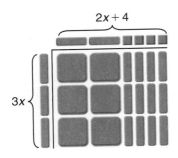

We use six x^2-tiles and twelve x-tiles.
So, $3x(2x + 4) = 6x^2 + 12x$

Use an area model

Sketch a rectangle with width $3x$ and length $2x + 4$.
The area of the rectangle is: $3x(2x + 4)$
Divide the rectangle into 2 rectangles.

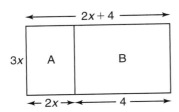

Rectangle A has area: $(3x)(2x) = 6x^2$
Rectangle B has area: $(3x)(4) = 12x$
The total area is: $6x^2 + 12x$
So, $3x(2x + 4) = 6x^2 + 12x$

Use paper and pencil

$3x(2x + 4)$
Multiply each term in the brackets by the term outside the brackets.

$3x(2x + 4) = (3x)(2x) + (3x)(4)$
$\qquad\qquad\quad = 6x^2 + 12x$

Use the distributive law.

Practice

1. Write the product modelled by each set of tiles. Determine the product.

a)

b)

c)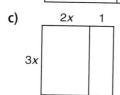

d)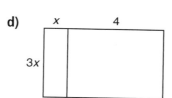

2. Write the product modelled by the area of each rectangle. Determine the product.

a)
	$3x$	4
$2x$		

b)
	$2x$	3
x		

c)
	$2x$	1
$3x$		

d)
	x	4
$3x$		

3. Use algebra tiles to expand.
  a) $x(3x + 2)$
  b) $x(4x + 6)$
  c) $3x(2x + 1)$
  d) $4x(3x + 4)$

4. Use an area model to expand.
  a) $4x(x + 3)$
  b) $x(2x + 3)$
  c) $2x(x + 6)$
  d) $3x(5x + 2)$

5. Use the calculator screen at the right.
  a) What patterns do you see? Explain the patterns.
  b) Write the next 2 lines in the patterns.

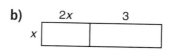

```
F1▾  F2▾  F3▾  F4▾  F5   F6▾
Tools Algebra Calc Other PrgmIO Clean Up
■ expand(x·(x + 3))      x² + 3·x
■ expand(x·(x + 4))      x² + 4·x
■ expand(x·(x + 5))      x² + 5·x
■ expand(x·(x + 6))      x² + 6·x
expand(x*(x+6))
MAIN        RAD AUTO      FUNC      4/30
```

6. **Assessment Focus** Alex thinks that $x(x + 1)$ simplifies to $x^2 + 1$.
   Choose a tool.
   Use the tool to explain how to get the correct answer.

When we cannot use area models, we use paper and pencil.

> **Example**
>
> Expand: $-2x(4x^2 + 2x - 3)$
>
> **Solution** Multiply each term of the polynomial by the monomial.
>
> $-2x(4x^2 + 2x - 3)$
> $= (-2x)(4x^2) + (-2x)(2x) + (-2x)(-3)$
> $= -8x^3 + (-4x^2) + (+6x)$
> $= -8x^3 - 4x^2 + 6x$

7. Expand.
   a) $2x(6x - 2)$
   b) $x(8x - 2)$
   c) $5x(x - 7)$
   d) $x(-4x + 3)$
   e) $2x(3x^2 - 4)$
   f) $4x(-2x - 3)$
   g) $3x(2x - 4)$
   h) $x(-x - 1)$

8. Expand.
   a) $-2x(x + 3)$
   b) $-x(2x + 1)$
   c) $-x(-2x - 7)$
   d) $-3x(2 - x)$
   e) $-7x(4x - 9)$
   f) $-x(2x + 3)$
   g) $-2x(3x - 5)$
   h) $-6x(2x + 3)$

9. Expand. Which method did you use each time?
   a) $2x(x^2 - 2x)$
   b) $-3x(3x^2 - 5x)$
   c) $4x(8x^2 + 6)$
   d) $7x(2x^2 + 3x - 1)$
   e) $2x(-4x^2 + 5x - 7)$
   f) $3x(4x^2 - 3x - 9)$
   g) $-2x(x - 3x^2 + 1)$
   h) $x(x^2 - 9 + x)$

10. **Take It Further**
    This figure shows one rectangle inside another.
    Determine an expression for the shaded area.
    Simplify the expression.

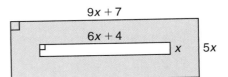

In Your Own Words

How is multiplying a polynomial by a monomial like multiplying a polynomial by a constant?
How is it different?
Use examples to explain.

7.7 Solving Equations with More than One Variable Term

Investigate Using An Equation to Solve a Problem

Jackie plans to start a business selling silk-screened T-shirts. She must purchase some equipment at a cost of $3500. Jackie will spend $6 per shirt on materials.

Jackie's total cost, in dollars, can be represented by the expression $3500 + 6x$, where x is the number of T-shirts she makes.
Jackie sells each shirt for $11.
Her income in dollars is represented by the expression $11x$.

When Jackie's income is equal to her total cost, she breaks even. This situation can be modelled by the equation:
$11x = 3500 + 6x$

Which tools could you use to help you?

➤ Solve the equation.
  How many T-shirts does Jackie need to sell to break even?

Reflect

➤ What strategies did you use to solve the equation?

➤ If you did not use CAS, explain how you could use it to solve the equation.

➤ Suppose Jackie sold each T-shirt for $13.
  What would the equation be?
  What would the solution be?

Connect the Ideas

We can use balance scales to model the steps to solve an equation.

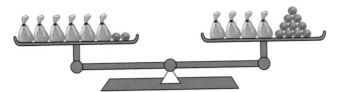

Each bag contains the same number of candies.

Let x represent the number of candies in a bag.
There are 6 bags and 2 candies on the left pan.
This is represented by the expression $6x + 2$.
There are 4 bags and 10 candies on the right pan.
This is represented by the expression $4x + 10$.
So, the balance scales model the equation: $6x + 2 = 4x + 10$

To solve the equation, remove
4 bags from each pan.
We now have: $2x + 2 = 10$

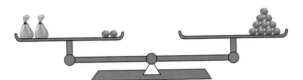

Remove 2 candies from each pan.
We now have: $2x = 8$

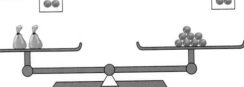

Divide the candies on the right pan
into 2 equal groups.
Each group contains 4 candies.

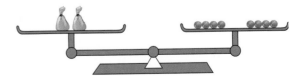

So, each bag contains 4 candies.
The solution is: $x = 4$

284 CHAPTER 7: Polynomials

Practice

1. Write the equation represented by each balance scales.
   Solve the equation.
   Tell how you did it.

   a)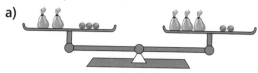

   b)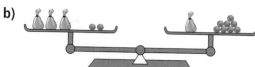

   c)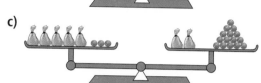

   d)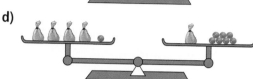

You can use a balance strategy to solve equations with pencil and paper.

Example

Solve: $3 + x = -4x - 42$
Check the solution.

Solution $3 + x = -4x - 42$

There are variables on both sides of the equals sign.
Perform the same operations on each side until the variables are on one side of the equals sign.

When we add the opposite term, we use inverse operations.

$3 + x = -4x - 42$	To remove $-4x$, add $4x$ to each side.
$3 + x + 4x = -4x - 42 + 4x$	Simplify.
$3 + 5x = -42$	To remove 3, add -3 to each side.
$3 + 5x - 3 = -42 - 3$	
$5x = -45$	To isolate x, divide by 5.
$\frac{5x}{5} = \frac{-45}{5}$	
$x = -9$	

Check the solution.
Substitute $x = -9$ in $3 + x = -4x - 42$.

L.S. $= 3 + x$ R.S. $= -4x - 42$
  $= 3 + (-9)$ $= -4(-9) - 42$
  $= 3 - 9$ $= 36 - 42$
  $= -6$ $= -6$

The left side equals the right side.
So, the solution is correct.

2. Solve. Explain your steps.
   a) $7 + x = 2x$
   b) $3x + 4 = 2x$
   c) $6 + 2x = x$
   d) $6x - 2 = 7x$
   e) $4x = 7 - 3x$
   f) $3x = 2x + 8$

3. Solve.
   a) $3x + 4 = 2x - 3$
   b) $-5 + 9x = 11 + 5x$
   c) $2 + 7x = 2x - 3$
   d) $x + 12 = 30 - 2x$
   e) $5 - 7x = -3x + 9$
   f) $-11 + 6x = -6x + 13$
   g) $-4x + 12 = 2x + 18$
   h) $50 + 7x = 8x + 1$

4. Choose two of the equations you solved in question 3. Check your solution.

5. Solve each equation.
   a) $2x + 12 = x + 20$
   b) $-6x + 15 = -x + 5$
   c) $-12 + 18x = 3x + 3$
   d) $-2x + 3x = 4x + 9$
   e) $5 - 6x - 12 = 14 - 7x$
   f) $4x - 5x + 7 = 2x - 14$

6. **Assessment Focus** Solve each equation. Show your work.
   a) $3x - 5 = 7 - 3x$
   b) $12 + 3x = x - 14$
   c) $-6x - 10 = 3x + 8$
   d) $-x = x + 6$
   e) $9 - 6x = x + 2$
   f) $8x - 4 - 3x = 11 + 4x$

7. An auto parts manufacturer buys a machine to produce a specific part.
   The machine costs $15 000.
   The cost to produce each part is $2.
   The parts will sell for $5 each.
   Let x represent the number of parts produced and sold.
   To break even, the cost, in dollars, 15 000 + 2x must equal the income 5x.
   This can be modelled by the equation: $15\,000 + 2x = 5x$
   Solve the equation.
   What does the solution represent?

8. **Take It Further** Solve each equation.
   a) $4(x - 3) = 3x$
   b) $12(5 - x) = 72$
   c) $-3(2x - 5) = -x + 5$

In Your Own Words

How many different ways can you solve an equation?
Use an example to illustrate your answer.

Chapter Review

What Do I Need to Know?

Polynomial

A polynomial is the sum or difference of one or more terms.

Algebra Tiles

We can represent a polynomial with algebra tiles.

 $-2x^2 + 4x + 3$

Monomial

A monomial is a polynomial with one term.
2, $3x$, $-4x^2$, and $5x^3$ are examples of monomials.

Constant Term

A constant term is a term without a variable; for example, 5.

Like Terms

Like terms are represented by the same type of algebra tile.
These are like terms:

$3x^2$ and $-2x^2$ $-x$ and $2x$ -3 and 2

The Distributive Law

Each term in the brackets is multiplied by the term outside the brackets.

$-2(-4x^2 + 3x - 2) = 8x^2 - 6x + 4$

Solving an Equation

To solve an equation means to determine the value of the variable that makes the equation true.

What Should I Be Able to Do?

7.1

1. Explain why x^2 and $2x$ are not like terms.

2. Which expression does each group of tiles represent?
   a)
   b)
   c)

3. Use algebra tiles to show each expression.
   Sketch the tiles you used.
   a) $3x^2 + 4x - 2$
   b) $-x^2 - 6x$

4. Simplify.
   a) $5 + 7x + 2$
   b) $4x + 6 + 3x$
   c) $2x^2 - 8x + x^2 + 3x$
   d) $6 + 3x - 3 - 2x$
   e) $2x^3 - 3 + 6 - 5x^3$
   f) $-2x^3 - x + 3x^3 + 5x$

5. Write an expression with 5 terms that has only 3 terms when simplified.

7.2

6. Add. Use algebra tiles.
   a) $(4x + 3) + (3x - 5)$
   b) $(x^2 + 6x) + (-3x^2 - 7x)$
   c) $(3x^2 + 1) + (2x^2 - 3)$
   d) $(x + 4) + (2x + 7)$
   e) $(3x^2 - 2x + 5) + (4x^2 + 5x - 7)$
   f) $(5x^2 + 7x - 3) + (-2x^2 - 3x + 2)$
   g) $(x^2 - 3x + 4) + (3x^2 - 4x - 3)$

7. Add. Which tools can you use?
   a) $(6x - 2) + (3x - 1)$
   b) $(2x^2 + 6x) + (-3x^2 - 2x)$
   c) $(3x^2 - 6x + 8) + (-5x^2 - x + 4)$
   d) $(-x^2 - 2x + 8) + (4x^2 - x + 3)$

7.3

8. Subtract. How could you check your answers?
   a) $(x - 8) - (4 + x)$
   b) $(x^2 + 4) - (3x^2 - 5)$
   c) $(7x^2 - 2x) - (5x^2 + 3x)$
   d) $(4x^2 - 3x - 1) - (2x^2 - 5x - 3)$
   e) $(-6x^2 + 5x + 1) - (2x^2 - x - 4)$
   f) $(2x^2 - 7x + 9) - (x^2 - 3x - 3)$

7.4

9. Expand. *Which tools could you use to help you?*
   a) $5(3x^2 + 4)$
   b) $4(7 + 2x)$
   c) $3(x^3 - 2x^2 + 2)$
   d) $8(2x^2 + 3x - 4)$
   e) $7(2x^3 - 1)$

10. Expand.
    a) $-6(x + 2)$
    b) $-7(x^2 - 4)$
    c) $-2(2x^2 + 1)$
    d) $-5(3x^3 - 2x - 3)$
    e) $-3(x^3 - 2x^2 + 4)$

11. Write the next 3 lines in the pattern shown on the calculator screen.

    ```
    expand(4·(x − 1))      4·x − 4
    expand(4·(2·x − 1))    8·x − 4
    expand(4·(3·x − 1))    12·x − 4
    expand(4*(3x−1))
    ```

7.5 12. Use algebra tiles to explain why:
  a) $3x + 2x$ equals $5x$
  b) $(3x)(2x)$ equals $6x^2$

13. a) Simplify.
   i) $(8x)(3x)$
   ii) $(-2x)(-6x)$
   iii) $(4x^2)(-3x)$
   iv) $(-9x)(-2x^2)$
  b) For which products in part a can you use algebra tiles? Explain.

7.6 14. Expand. For which products can you use algebra tiles? Explain.
  a) $x(3x + 4)$
  b) $3x(x^2 - 8)$
  c) $2x(4 - x)$
  d) $6x(-3x^2 + 4x + 2)$

15. Expand.
  a) $-3x(-2x + 3)$
  b) $-2x(x^2 - 5)$
  c) $-5x(3x + 7)$
  d) $2x(4x^2 + 5x - 3)$

16. Write the next 3 lines in the pattern shown on each screen.
  a)
  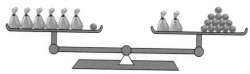
  b)
  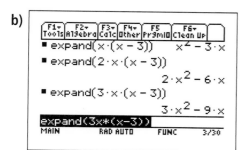

7.7 17. a) Let x represent the number of candies in each bag. Write the equation represented by the scales.

  b) Solve the equation.
  c) How many candies are in each bag?

18. Solve each equation.
  a) $5x + 8 = x$
  b) $3x + 3 = x + 7$
  c) $2x + 10 = 4$

19. Solve each equation.
  a) $5x - 4 = 8 + 3x$
  b) $6 + 3x = x - 2$
  c) $12 + x = -2x + 9$
  d) $2x - 3 = 6 - x$

20. A fund-raiser is organized for hurricane victims. With the purchase of a $100 ticket, each person is given a souvenir bracelet (value $20) and the chance to win a car.
  Let x represent the number of tickets sold.
  Then, the income, in dollars, from ticket sales is $100x$.
  The expenses, in dollars, are $20\,000 + 20x$.
  The organizers of the fundraiser would like to raise $60 000 after all expenses. This can be modelled by the equation:
  $100x = 60\,000 + 20\,000 + 20x$
  a) Solve the equation.
  b) What does the solution represent?

Chapter Review **289**

Practice Test

Multiple Choice: Choose the correct answer for questions 1 and 2.

1. Which polynomial is simplified?
   A. $3x + 4 - x^2 + 8$
   B. $3x^3 - 2x + x^2 - x$
   C. $x^2 - 6 + x$
   D. $x + 6x - x^2 + 7$

2. What is the solution to $80 + 10x = 30x - 20$?
   A. $x = 3$
   B. $x = 3.5$
   C. $x = 5$
   D. $x = -5$

Show your work for questions 3 to 6.

3. **Knowledge and Understanding** Simplify.
   a) $(3x^2 + 4x - 1) + (2x^2 - 8x - 4)$
   b) $(x^2 + 3x - 2) - (2x^2 + x - 2)$
   c) $3(x + 4)$
   d) $(2x)(3x^2)$
   e) $4x(x^2 - 5x + 3)$

   Which tools could you use to help?

4. **Application** The cost to rent a hall for the prom is $400 for the hall and $30 per person for the meal. This can be modelled by the equation $C = 400 + 30x$, where x is the number of students attending.
   a) Suppose 150 students attend. What will be the cost of the prom?
   b) The prom committee has $10 000. What is the greatest number of students that can attend with this budget?

5. **Communication** How can you tell if a polynomial can be simplified?
   Include examples in your explanation.

6. **Thinking** Joe subtracted $(4x^2 - 3x) - (2x^2 - 5x + 4)$.
   He got the answer $2x^2 - 8x + 4$.
   a) What mistake did Joe likely make? Explain.
   b) Determine the correct answer.
   c) How could you check your answer is correct?

Chapters 1–7 Cumulative Review

CHAPTER 1

1. Determine the perimeter and area of each figure. The curve is a semicircle.
   a)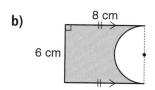
   b)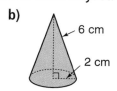

2. Determine the volume of each object.
   a) b)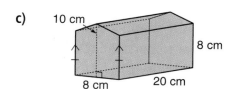
   c)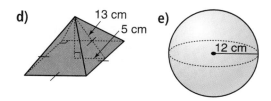
   d) e)

CHAPTER 2

3. a) For each perimeter, determine the dimensions of the rectangle with the maximum area.
   i) 40 cm   ii) 68 m   iii) 90 cm
   b) Calculate the area of each rectangle in part a. How do you know each area is a maximum?

4. Determine the dimensions of a rectangle with area 144 cm² and the minimum perimeter. What is the minimum perimeter?

5. A patio is to be built on the side of a house using 48 congruent square stones. It will then be surrounded by edging on the 3 sides not touching the house. Which designs require the minimum amount of edging? Include a labelled sketch in your answer.

CHAPTER 3

6. Can a right triangle have an obtuse angle? Explain.

7. Determine the angle measure indicated by each letter. Justify your answers.
   a)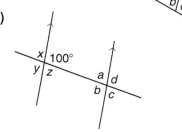
   b)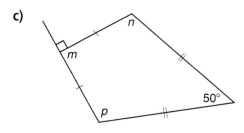
   c)

8. Determine the measure of one exterior angle of a regular polygon with each number of sides. Show your work.
   a) 5 sides   b) 9 sides   c) 10 sides

Cumulative Review **291**

CHAPTER 4

9. Determine the value of each variable.
   a) $2:5 = 6:n$
   b) $3:b = 7:42$
   c) $a:15 = 4:10$
   d) $9:12 = 15:m$

10. Determine each unit rate.
    a) 288 km driven in 4 h
    b) $3.79 for 3 kg of apples
    c) $51 earned for 6 h of work
    d) 9 km hiked in 2.5 h
    e) $3.49 for 6 muffins

11. Maria is swimming lengths. She swims 2 lengths in 1.5 min. At this rate:
    a) How far can she swim in 9 min?
    b) How long will it take her to swim 30 lengths?

12. A bathing suit is regularly priced at $34.99. It is on sale at 30% off.
    a) What is the sale price?
    b) How much does the customer pay, including taxes?

13. Tomas borrows $2500 for 6 months. The annual interest rate was 3%.
    a) How much simple interest does Tomas pay?
    b) What does the loan cost Tomas?

CHAPTER 5

14. Your teacher will give you a large copy of the scatter plot below.

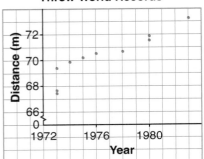
Women's Discus Throw World Records

a) What does the scatter plot show?
b) What was the world record in women's discus in 1975? In 1980?
c) In what year was the world record about 70 m?
d) Draw a line of best fit.
e) Estimate the world record in 1984. Explain how you did this.

15. Here are the dimensions of some rectangles with perimeter 30 cm.

Width (cm)	1	2	3	4	5	6	7
Length (cm)	14	13	12	11	10	9	8

a) Is the relationship between length and width linear? Justify your answer.
b) Graph the data. Does the graph illustrate your answer to part a? Explain.
c) Use the graph.
   i) Determine the length when the width is 5.5 cm.
   ii) Determine the width when the length is 13.5 cm.
d) Write a rule for the relationship.

16. A baseball is thrown up into the air. Its height is measured every 0.2 s.

Time (s)	0	0.2	0.4	0.6	0.8	1.0	1.2
Height (m)	1.0	2.1	2.8	3.0	2.9	2.5	1.6

a) Graph the data.
b) Draw a curve or line of best fit.
c) What is the greatest height the ball reaches?
d) When will the ball hit the ground?
e) At what times is the ball 2 m above the ground?

CHAPTER

17. The graph shows Tyler's distance from his home as he drives to his cottage.

Tyler's Drive

a) Describe Tyler's drive.
b) Tyler makes 2 stops: one to buy gas and another to pick up his cousin. Which part of the graph do you think represents each stop? Justify your answers.

18. This pattern is made of toothpicks. It continues.

Frame 1   Frame 2   Frame 3

a) Sketch the next frame in the pattern.
b) Copy and complete the table below for the first 6 frames. Is the relation linear or non-linear? Explain.

Frame number	Number of toothpicks	First differences
1		
2		
3		

c) Graph the relationship. Does the graph support your answer to part b? Explain.

19. The table shows the value in Philippine pesos of different amounts in Canadian dollars in early 2006.

Amount in dollars	10	55	48	20	36
Amount in pesos	450	2475	2160	900	1620

a) Graph the data. Does the graph represent direct variation? Explain.
b) Determine the rate of change. Explain what it represents.
c) Write an equation for the value, p pesos, of d dollars.
d) Determine the value of $175.
e) Determine the value of 7500 pesos.
f) How did you answer parts d and e? Did you use the table, graph, or equation? Explain.

20. Refer to question 19. In early 1999, $1 was worth about 25 Philippine pesos.
a) How would the graph change? Draw the new graph on the grid in question 19.
b) How would the equation change? Write the new equation.

21. Sunil is joining a tennis club for the summer. The club offers a special 10-week summer membership for students. It costs $50, plus $3.50 per hour of court time.
a) Make a table. Show the total costs for times played from 0 h to 60 h.

Cumulative Review **293**

b) Graph the data. Does the graph represent direct variation? Explain.
c) Write an equation to determine the total cost, C dollars, when n hours of tennis are played.
d) Suppose Sunil has budgeted $200 for tennis costs. How many hours can he play? How did you determine your answer?

22. Refer to question 21. The tennis club also offers a special 10-week summer membership for adults. It costs $75, plus $4.00 per hour of court time.
a) How would the graph change? Draw the new graph on the grid in question 21.
b) How would the total cost equation change? Write the new equation.
c) How many hours could an adult play if her budget for tennis is $200?

23. The daily cost of running a sausage cart is a fixed cost of $40, plus $1 per sausage. The revenue is $3 per sausage.
a) Write an equation for the daily cost, C dollars, in terms of n, the number of sausages sold.
b) Write an equation for the revenue, R dollars, in terms of n.
c) Graph the equations on the same grid.
d) How many sausages have to be sold before a profit is made? Explain.

24. What expression does each group of algebra tiles represent?
a)
b)

25. Simplify each expression by combining like terms.
a) $8x + 3 - x$
b) $3x^2 - 7x - x^2 + 4x$
c) $10 + 2x - 5 - 2x$
d) $3x^3 + 6x + 2x^3 - 2x$

26. Add. Which tools can you use?
a) $(5x + 3) + (6x - 7)$
b) $(2x^2 - 9x) + (4x - 3x^2)$
c) $(-x^2 + 3x - 1) + (3x^2 - 7x + 2)$

27. Subtract. How could you check your answers?
a) $(2x + 5) - (x + 2)$
b) $(7x^2 - 8) - (6x^2 - 3)$
c) $(5x^2 + x - 2) - (-2x^2 + 3x - 3)$

28. Expand.
a) $6(5 + 2x)$
b) $2(x^3 - 3x^2 + 3)$
c) $-3(3x + 1)$
d) $-4(2x^3 - 5x^2)$
e) $x(2x + 7)$
f) $3x(7 - 3x)$
g) $-2x(2x^2 - 3)$
h) $2x(3x^2 - 4x + 5)$

29. Solve each equation. How could you check your answer?
a) $7x + 6 = 10 + 3x$
b) $8 + 5x = x - 4$
c) $9 - 3x = -5x + 3$
d) $2x - 2 = 4 - x$

Extended Glossary

acute angle: an angle whose measure is less than 90°

acute triangle: a triangle with all angles less than 90°

alternate angles: two angles that are between two lines, but are on opposite sides of a transversal that intersects the two lines

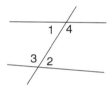

Angles 1 and 2 are alternate angles.
Angles 3 and 4 are alternate angles.

approximation: a number close to the exact value of a quantity or an expression; the symbol $\doteq$ means "is approximately equal to"

area: the number of square units needed to cover a surface; common units used to measure area include square centimetres and square metres

average speed: the speed that, if the object travelled at that speed constantly, would result in the same total distance being travelled in the same total time; to calculate average speed, the total distance travelled during the given time period is divided by the total time

In 4 h, a car travels 285 km.
Determine the average speed of the car.
$$\text{Average speed} = \frac{\text{Distance}}{\text{Time}}$$
$$= \frac{285 \text{ km}}{4 \text{ h}}$$
$$= 71.25 \text{ km/h}$$

base: the side of a polygon, or the face of a solid or object, from which the height is measured

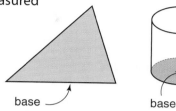

binomial: a polynomial with two terms

$3x + 8$ and $4x^3 - 7$ are binomials.

capacity: the amount a container can hold; common units used to measure capacity include millilitres and litres

circle: the set of points in a plane that are a given distance (the radius) from a fixed point (the centre)

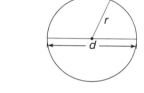

The area of a circle is:
$A = \pi r^2$, where r is the radius
The circumference of a circle is:
$C = 2\pi r$, where r is the radius,
or $C = \pi d$, where d is the diameter
The circumference of a circle is also the perimeter of the circle.
The area of a circle with radius 5 cm is:
$A = \pi(5)^2$
$= \pi(25)$
$\doteq 79$

The area is approximately 79 cm².
The circumference of a circle with radius 5 cm is:
$C = 2\pi(5)$
$\doteq 31$
The circumference is approximately 31 cm.

EXTENDED GLOSSARY

circumference: the distance around a circle

coefficient: the numerical factor of a term

common factor: a number that is a factor of each of the given numbers

3 is a common factor of 9, 12, and 30.

composite figure: a figure that is made up of other, simpler figures

cone: a solid that is formed by a region (the *base* of the cone) and all the line segments joining points in the base to a point not in the base

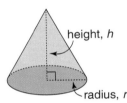

The volume of a cone is one-third the volume of the related cylinder and can be found using the formula:
$V = \frac{1}{3}$(Base area)(Height)
When the circular base has radius r, and the height of the cone is h, this formula becomes:
$V = \frac{1}{3}(\pi r^2)(h)$
The volume of a cone whose circular base has radius 6 cm and with height 14 cm is:
$V = \frac{1}{3} \times \pi \times (6)^2 \times (14)$
$\doteq 528$

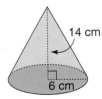

The volume is approximately 528 cm³.

congruent: figures that have the same size and shape, but not necessarily the same orientation

constant term: a number

coordinates: the numbers in an ordered pair that locate a point on a grid

corresponding angles: two angles that are on the same side of a transversal that intersects two lines and on the same side of each line

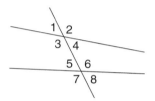

Angles 1 and 5 are corresponding angles.
Angles 2 and 6 are corresponding angles.
Angles 3 and 7 are corresponding angles.
Angles 4 and 8 are corresponding angles.

cube: a rectangular solid whose length, width, and height are all equal

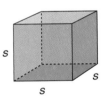

The volume of a cube is:
$V = s^3$, where s is the edge length
The volume of a cube with edge length 7 cm is:
$V = (7)^3$
$= 343$

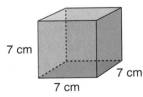

The volume is 343 cm³.

curve of best fit: a curve that passes as close as possible to a set of plotted points

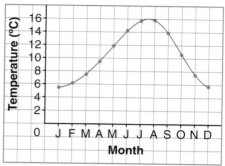

cylinder: a solid with two parallel, congruent, circular bases

The volume of a cylinder can be found using the formula:
$V = $ (Base area)(Height)
$V = (\pi r^2)(h)$,
where the circular base has radius r, and the height of the cylinder is h
The volume of a cylinder whose circular base has radius 8 cm and with height 20 cm is:
$V = \pi \times (8)^2 \times 20$
$\doteq 4021$

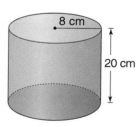

The volume is approximately 4021 cm³.

diagonal: a line segment that joins two vertices of a figure, but is not a side

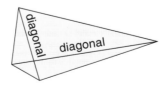

diameter: a line segment that joins two points on a circle (or surface of a sphere) and passes through its centre; the diameter of a circle (or sphere) is twice the length of the radius

direct variation: when one quantity is a constant multiple of another quantity; that is, the quantities are proportional; the graph that represents direct variation is a straight line that passes through the origin

Pierre has a summer job planting trees. He earns 17¢ for each tree he plants. This situation represents direct variation. To determine Pierre's earnings in dollars, we multiply the number of trees planted by 0.17.

A table and graph for this relation are shown.

Trees planted	Earnings ($)
0	0
100	17.00
200	34.00
300	51.00
400	68.00
500	85.00

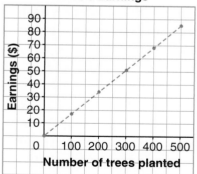

distributive law: the property stating that a product can be written as a sum or difference of two products; for example:

$a(b + c) = ab + ac$ and
$a(b - c) = ab - ac$

EXTENDED GLOSSARY **297**

equation: a mathematical statement indicating that two expressions are equal

equation of a line: an equation for the relationship between the coordinates of every point on a line

equilateral triangle: a triangle with three equal sides; each angle is 60°

equivalent: having the same value

$\frac{2}{3}$ and $\frac{6}{9}$ are equivalent fractions.
2 : 3 and 6 : 9 are equivalent ratios.

equivalent expressions: numerical expressions that have the same value; if the expressions contain variables, expressions that result in the same value for all possible values of the variable

$3(2x + 3)$ and $6x + 9$ are equivalent expressions.

estimate: a reasoned guess that is close to the actual value, without calculating it exactly

evaluate an expression: substitute a number for each variable in the expression, then work out the resulting arithmetic expression applying the order of operations rules

Evaluate $2x^2 + 3x - 5$, when $x = -2$.
Replace each x with -2, placing each number in brackets to prevent errors with signs.
$2x^2 + 3x - 5 = 2(-2)^2 + 3(-2) - 5$
$= 2(4) + 3(-2) - 5$
$= 8 - 6 - 5$
$= -3$

expand an expression: use the distributive law to multiply parts of an expression

Expand $2(3x^2 - 2x + 7)$.
Use the distributive law, using brackets to prevent errors with signs.
$2(3x^2 - 2x + 7)$
$= 2(3x^2) + 2(-2x) + 2(7)$
$= 6x^2 - 4x + 14$

exponent: a number, placed at the right of and above another number or expression, that tells how many times the number or expression before it is used as a factor

3 is the exponent in 6^3.
6^3 means $6 \times 6 \times 6$.
2 is the exponent in x^2.
x^2 means $(x)(x)$.

expression: a mathematical phrase made up of numbers and/or variables connected by operations

exterior angle: the angle between one side of a polygon and the extension of an adjacent side of the polygon

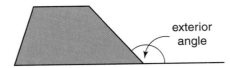

factor: any integer that divides exactly into a given integer; to factor an integer means to write it as a product of integers

The factors of 20 are 1, 2, 4, 5, 10, and 20, because $1 \times 20 = 2 \times 10 = 4 \times 5 = 20$.

first differences: when data are arranged in a table, with the data in the first column increasing in constant steps, the first differences are found by subtracting consecutive numbers in the second column; if all the first differences are equal, then the relationship is linear

Consider the relationship between the time worked and the money earned.

Time worked (h)	Money earned ($)	Change in money earned ($)
0	0	
1	9	$9 - 0 = 9$
2	18	$18 - 9 = 9$
3	27	$27 - 18 = 9$
4	36	$36 - 27 = 9$

All the first differences are 9.
Since they are constant, we know the relationship is linear.

fixed cost: a cost that remains constant

formula: an equation that describes the relationship between two or more quantities

> The formula that describes how the area, A, of a rectangle is related to its length, ℓ, and width, w, is
> $A = \ell \times w$, or
> $A = \ell w$

hexagon: a polygon with six sides

horizontal axis: the horizontal number line on a coordinate grid

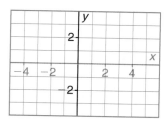

hypotenuse: the side opposite the right angle in a right triangle

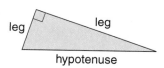

integers: the set of numbers …−3, −2, −1, 0, 1, 2, 3…

interest (simple): money paid for the use of money, usually at a predetermined percent. When P is the amount invested or borrowed, r the annual rate of interest written as a decimal, and t the time in years, then I, the interest, is given by the formula $I = Prt$

> Juanita purchased a $500 bond at an annual interest rate of 6%. After 6 months, she received the following interest:
> $I = Prt$
> $= 500 \times 0.06 \times \frac{6}{12}$
> $= 500 \times 0.06 \times 0.5$
> $= 15$
> The interest is $15.

interior angles: angles inside a polygon; angles that are between two lines and are on the same side of a transversal that intersects the lines

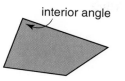 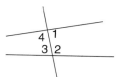

1 and 2 are interior angles.
3 and 4 are interior angles.

intersecting lines: lines that meet or cross; lines that have one point in common

inverse operation: an operation that reverses another operation

> Subtraction is the inverse of addition and division is the inverse of multiplication.

isosceles triangle: a triangle with two equal sides; the angles opposite the equal sides are also equal

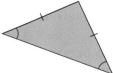

legs: the two shorter sides in a right triangle; see *hypotenuse*

like terms: terms that have the same variables with the same exponent

> $4x$ and $-3x$ are like terms.
> $4x$ and $-3x^2$ are not like terms.

line of best fit: a line that passes as close as possible to a set of plotted points

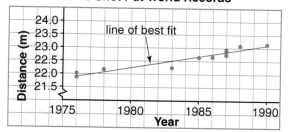

line segment: the part of a line between two points on the line, including the two points

EXTENDED GLOSSARY

linear relation (relationship): a relation (relationship) that can be represented by a straight line graph

mass: a measure of the amount of material in an object; common units used to measure mass are grams and kilograms

mean: one measure of the average of a set of numbers; to find the mean, add the numbers in the set then divide their sum by the number of terms in the set

> One week Nora walked these distances:
> 4 km, 5 km, 3 km, 4 km, 5 km, 4 km, 3 km.
> The total distance she walked is
> 4 km + 5 km + 3 km + 4 km + 5 km + 4 km + 3 km = 28 km
> There are 7 terms.
> So, the mean distance Nora walked each day is:
> $\frac{28 \text{ km}}{7} = 4$ km

median: one measure of average of a set of numbers; it is the middle number of a set of numbers arranged in numerical order; if there are two middle numbers, their mean is the median of the data set

> In the last 6 basketball games, Zack has scored these points:
> 12, 10, 14, 12, 9, 10
> Arrange from least to greatest:
> 9, 10, 10, 12, 12, 14
> There are 6 numbers in the set, so the median is the mean of the 3rd and 4th numbers.
> $\frac{10 + 12}{2} = \frac{22}{2}$
> $= 11$
> The median number of points Zack scored is 11.

monomial: a polynomial with one term

> 14 and $5x^2$ are monomials.

multiple: the product of a given number and an integer

> Some multiples of 8 are 8, 16, 24, 32, …

natural numbers: the set of numbers 1, 2, 3, …

non-linear relation (relationship): a relation (relationship) that cannot be represented by a straight line graph

obtuse angle: an angle greater than 90° and less than 180°

obtuse triangle: a triangle with one obtuse angle

octagon: a polygon with eight sides

opposite angles: the equal angles that are formed by two intersecting lines

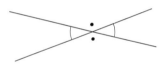

opposite integers: two integers with a sum of 0

> +3 and −3 are opposite integers.

opposite polynomials: two polynomials with a sum of 0

> $2x^2 − 5x + 7$ and $−2x^2 + 5x − 7$ are opposite polynomials.

origin: the point (0, 0) on a graph; this is the point where the axes intersect

parallel lines: lines in the same plane that do not intersect

> Lines *m* and *n* are parallel. To show this, we draw a matching arrowhead on each line.

parallelogram: a quadrilateral with opposite sides parallel; the opposite sides are also equal

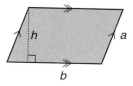

The perimeter of a parallelogram is:
$P = 2a + 2b$
The area is: $A = bh$
For the parallelogram below,

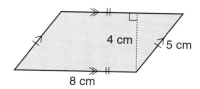

$P = 2 \times 8 + 2 \times 5$
$= 16 + 10$
$= 26$
The perimeter is 26 cm.
$A = 8 \times 4$
$= 32$
The area is 32 cm².

partial variation: when one quantity equals a fixed value plus a constant multiple of another quantity; the graph that represents partial variation is a straight line that does *not* pass through the origin

A taxi company charges a fixed cost of $2.75 plus $1.50 per kilometre.
To determine the total cost of a trip, we multiply the distance in kilometres by $1.50 and add the result to $2.75. A table and graph for this relation are shown.

Distance driven (km)	Cost ($)
0	2.75
2	5.75
4	8.75
6	11.75
8	14.75
10	17.75

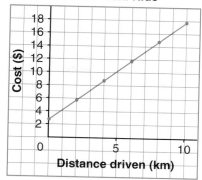

pentagon: a polygon with five sides

percent: means "out of 100"; it is a ratio that compares a number to 100

A percent can be written as a fraction with denominator 100, or as a decimal.
$45\% = \frac{45}{100} = 0.45$

A CD that sells for $17.99 is on sale for 25% off. What is the sale price?
The sale price is $100\% - 25\% = 75\%$ of the original price, or 75% of $17.99.
Let b dollars represent the sale price.
$b : 17.99 = 75 : 100$
Write this proportion in fraction form.
$\frac{b}{17.99} = \frac{75}{100}$
To isolate b, multiply each side of the equation by 17.99.
$\frac{b}{17.99} \times 17.99 = 0.75 \times 17.99$
$b = 13.4925$
Since prices are given in dollars and cents, the value of b is rounded to 2 decimal places. The sale price is $13.49.

perimeter: the distance around a closed figure; see *circle, parallelogram, rectangle, semicircle, square, trapezoid,* and *triangle*

perpendicular lines: lines that intersect at right angles (90°)

pi (π): the ratio of the circumference of a circle to its diameter

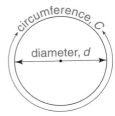

$C : d = \pi : 1$

polygon: a closed figure that consists of line segments that only intersect at their endpoints

These figures are polygons.

These figures are not polygons.

polynomial: a mathematical expression with one or more algebraic terms

$2x$, $3 - 5x$, and $x^2 + 2x - 8$ are polynomials.

power: a number with an exponent; see *exponent*

5^3 is a power of 5

prediction: a statement of what you think will happen

prism: a solid with two congruent and parallel faces (the *bases*); all other faces are parallelograms

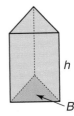

The volume of a prism can be found using the formula:
V = (Base area)(Height)
$= B \times h$, or Bh

For the prism below,

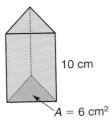

$V = 6 \times 10$
$= 60$
The volume is 60 cm³.

proportion: a statement that two ratios are equal; for example, $x : 12 = 2 : 5$

To solve for an unknown term in a proportion, write the ratios in fraction form, then multiply to isolate the term that contains x.

$$\frac{x}{12} = \frac{2}{5}$$
$$12 \times \frac{x}{12} = 12 \times \frac{2}{5}$$
$$x = \frac{12 \times 2}{5}$$
$$x = 4.8$$

pyramid: a solid with one face that is a polygon (*base*) and other faces that are triangles with a common vertex

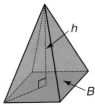

The volume of a pyramid is one-third the volume of the related prism and can be found using the formula:
$V = \frac{1}{3}$(Base area)(Height)

For the pyramid below,

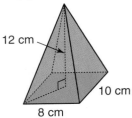

$V = \frac{1}{3}(8 \times 10)(12)$
$= 320$
The volume is 320 cm³.

Pythagorean Theorem: in any right triangle, the sum of the areas of the squares on the two shorter sides is equal to the area of the square on the hypotenuse.

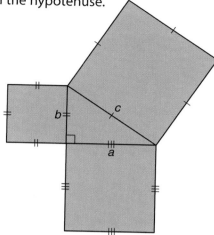

In a right triangle: $a^2 + b^2 = c^2$
For the right triangle below,

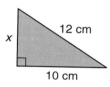

$10^2 + x^2 = 12^2$ Simplify each side.
$100 + x^2 = 144$ Isolate the variable.
$100 + x^2 - 100 = 144 - 100$
$x^2 = 44$ To determine x, take the square root of each side.
$x = \sqrt{44}$
$x \doteq 6.6$
The third side measures approximately 6.6 cm.

quadrant: one of the four regions into which the coordinate axes divide a grid

	y	
Second quadrant		First quadrant
	0	x
Third quadrant		Fourth quadrant

quadrilateral: a polygon with four sides

radius (*plural,* **radii**): the distance from the centre of a circle to any point on the circumference; also, the distance from the centre of a sphere to any point on the surface of the sphere

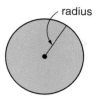

rate: a certain quantity considered in relation to another quantity

Speed is the rate at which distance changes in relation to time.

rate of change: a measure of how one quantity changes with respect to another; it can be determined by calculating $\frac{\text{rise}}{\text{run}}$

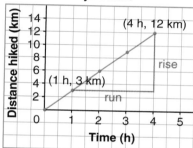

Choose 2 points on the line:
(1, 3) and (4, 12)
rise = 12 km − 3 km = 9 km
run = 4 h − 1 h = 3 h
Rate of change = $\frac{\text{rise}}{\text{run}}$
$= \frac{9 \text{ km}}{3 \text{ h}}$
$= 3$ km/h
The rate of change is 3 km/h.
This is Maya's average hiking speed.

ratio: a comparison of two quantities

Andrew is making orange punch for a party. He uses 2 cups of soda water for every 3 cups of orange juice. The ratio of soda water to orange juice is 2 : 3.

EXTENDED GLOSSARY

rectangle: a quadrilateral with four right angles

The perimeter of a rectangle is:
$P = 2\ell + 2w$
The area is: $A = \ell w$
For the rectangle below,

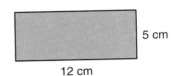

$P = 2 \times 12 + 2 \times 5$
$ = 24 + 10$
$ = 34$
The perimeter of the rectangle is 34 cm.
$A = 12 \times 5$
$ = 60$
The area is 60 cm².

rectangular prism: a prism with rectangular faces

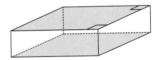

rectangular pyramid: a pyramid with a rectangular base

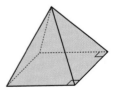

reflex angle: an angle greater than 180° and less than 360°

regular polygon: a polygon with all sides equal and all angles equal

The polygons below are regular polygons.

relation (relationship): a rule that explains how two quantities or measures are related

Nikhil enjoys cycling long distances. His distance travelled in kilometres is related to the cycling time by the rule: multiply the time in hours by 16. The relation can be represented by the equation $d = 16t$, where d is the distance in kilometres and t is the time in hours.

right angle: a 90° angle

right triangle: a triangle with one right angle

rise: the vertical distance between two points; see *rate of change*

run: the horizontal distance between two points; see *rate of change*

scale: the ratio of the distance between two points on a map, model, or diagram to the actual distance; also, the numbers labelling the coordinate axes on a grid

scale drawing: a drawing in which the lengths are an enlargement or a reduction of actual lengths

scalene triangle: a triangle with no equal sides

scatter plot: a graph of data that are a set of points

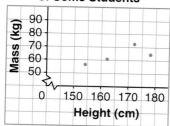

semicircle: half a circle

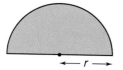

The area of a semicircle is:
$A = \frac{1}{2}\pi r^2$, where r is the radius
The perimeter of a semicircle is:
$P = \pi r + 2r$, where r is the radius

The area of a semicircle with radius 7 cm is:
$A = \frac{1}{2} \times \pi \times 7^2$
$= \frac{1}{2} \times \pi \times 49$
$\doteq 77$

The area is approximately 77 cm².
The perimeter of a semicircle with radius 7 cm is:
$P = \pi \times 7 + 2 \times 7$
$\doteq 36$
The perimeter is approximately 36 cm.

slant height: the distance from the top of a cone or pyramid to its base, measured along its sloped surface

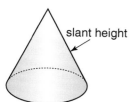 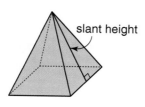

solving an equation: determining the value of the variable that makes the equation a true statement

Solve the equation $5x + 7 = 11 + 3x$.
Use inverse operations to collect like terms on one side of the equation.
Subtract $3x$ from each side, then simplify.
$5x + 7 - 3x = 11 + 3x - 3x$
$2x + 7 = 11$
Subtract 7 from each side, then simplify.
$2x + 7 - 7 = 11 - 7$
$2x = 4$
Divide each side by 2.
$\frac{2x}{2} = \frac{4}{2}$
$x = 2$

speed: see *average speed*

sphere: the set of points in space that are a given distance (the *radius*) from a fixed point (the *centre*)

The volume of a sphere is:
$V = \frac{4}{3}\pi r^3$, where r is the radius

The volume of a sphere with radius 7 cm is:
$V = \frac{4}{3} \times \pi \times (7)^3$
$= \frac{4}{3} \times \pi \times 343$
$\doteq 1437$

The volume of the sphere is approximately 1437 cm³.

square: a rectangle with four equal sides

The perimeter of a square is: $P = 4s$
The area is: $A = s^2$

For the square below,

$P = 4 \times 11$
$= 44$

The perimeter is 44 cm.

$A = 11^2$
$= 121$

The area is 121 cm².

square root: a number which, when multiplied by itself, results in a given number

5 and -5 are the square roots of 25, since $5^2 = 25$ and $(-5)^2 = 25$. The notation $\sqrt{25}$ represents the positive square root only.

straight angle: a 180° angle

substituting into an equation: in an equation, replacing a variable with a number then solving the equation for the remaining variable

Annie's earnings, E dollars, are given by the equation $E = 8.5n$, where n is the time in hours she works.
To find the time Annie would have to work to earn $150, substitute $E = 150$ then solve the equation for n.

$150 = 8.5n$ Divide each side by 8.5.
$\frac{150}{8.5} = \frac{8.5n}{8.5}$
$17.647 \doteq n$

Annie would have to work about 18 h to earn $150.

term: when an algebraic expression is written as the sum of several quantities, each quantity is a term of the expression

transversal: a line that intersects two or more lines

Line t is a transversal.

trapezoid: a quadrilateral with one pair of parallel sides

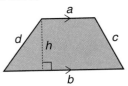

The perimeter of a trapezoid is:
$P = a + b + c + d$
The area of a trapezoid is:
$A = \frac{1}{2}(a + b)h$

For the trapezoid below,

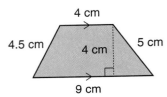

$P = 4 + 5 + 9 + 4.5$
$= 22.5$

The perimeter is 22.5 cm.

$A = \frac{1}{2}(4 + 9)(4)$
$= \frac{1}{2}(52)$
$= 26$

The area is 26 cm².

trend: a relationship between measures that is shown by a graph of data

triangle: a polygon with three sides

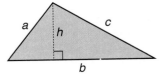

The perimeter of a triangle is:
$P = a + b + c$
The area of a triangle is:
Area $= \frac{1}{2}$(Base)(Height)
$A = \frac{1}{2}bh$

For the triangle shown below,

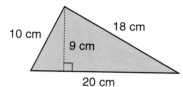

$P = 10 + 18 + 20$
$= 48$
The perimeter is 48 cm.
$A = \frac{1}{2}(20 \times 9)$
$= 90$
The area is 90 cm².

triangular prism: a prism with triangular bases

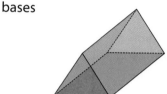

triangular pyramid: a pyramid with a triangular base

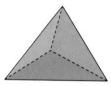

unit price: the price of one item or the price of a particular mass or volume of an item

$1.15 per litre is a unit price.
It is written as $1.15/L.

unit rate: the quantity associated with a single unit of another quantity

6 m in 1 s is a unit rate.
It is written as 6 m/s.

variable: a letter or symbol used to represent a quantity that varies

vertex (*plural,* **vertices**): the corner of a figure or solid

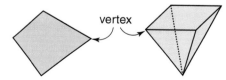

vertical axis: the vertical number line on a coordinate grid

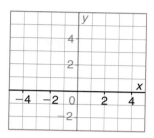

vertical intercept: the vertical coordinate of the point at which the graph of a line intersects the vertical axis

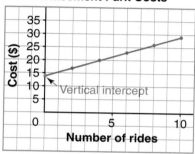

volume: the amount of space occupied by an object; common units used to measure volume include cubic metres and cubic centimetres

whole numbers: the set of numbers 0, 1, 2, 3, …

zero pair: two opposite integers whose sum is 0; two opposite algebra tiles whose sum is 0

Index

A
acute triangle, 76
algebra,
    solving proportions with, 131, 132, 133
algebra tiles, 253, 254, 258, 263, 264, 269, 270, 275, 276, 279, 280
algebraic expressions, 253, 254
alternate angles, 86, 88
angles,
    in quadrilaterals, 95, 96
    in triangles, 75, 76, 78, 81, 82, 84
    investigating with *The Geometer's Sketchpad*, 79, 80, 85, 86, 93, 94
    involving parallel lines, 87, 88, 89, 97
    on a straight line, 88, 96
area (*see also* maximum area), 3, 4
    investigating with *The Geometer's Sketchpad*, 48
    maximum for a given perimeter, 53, 54, 55, 66
    of a circle, 4
    of a parallelogram, 4
    of a semicircle, 13
    of a trapezoid, 4
    of a rectangle, 4, 46
    of composite figures, 11, 12
average speed, 197, 198, 199, 239, 242

B
balance scales models, 228, 284

C
capacity, 22
circle,
    area of, 4
    circumference of, 3, 4, 16
coefficient, 254, 256, 276, 278
common factor, 112
composite figure,
    area of, 11, 12
    perimeter of, 15, 16
cone,
    volume of, 29, 30
constant term, 270
coordinates, 246
corresponding angles, 86, 88
curve of best fit, 159, 160, 162, 176, 178
    drawing with a graphing calculator, 164–166
cylinder,
    volume of, 21, 22, 29, 30, 34, 35

D
direct variation, 205, 206, 208, 232
    change in direct variation, 221, 222
distance-time graph, 198, 199, 239, 242
distributive law, 270

E
equal angles, 172
equations,
    developing from a graph, 235–237, 239
    for a graph, 206, 209, 213, 215, 223, 232, 233
    of a line, 231
    solving, 227, 228, 229
    with more than one variable term, 283, 284, 285
equilateral triangle, 103
equivalent ratios, 111, 112, 116, 117
expanding, 270
exponential curve of best fit, 164, 165
exponents, 254
exterior angles, 81, 82, 84
    of polygons, 99, 100
    of quadrilaterals, 95, 96
    investigating with *The Geometer's Sketchpad*, 80, 94, 104

F
Fathom,
    drawing a line of best fit, 156–158
first differences, 193, 195
Frayer model, 74

G
Games:
    Find the Least Perimeter, 59
    Hidden Sum, 167
    I Have…Who Has…?, 267
    Measurement Bingo, 19
    The 25-m Sprint, 219
Geometer's Sketchpad, The
    investigating angles in polygons, 103, 104
    investigating angles in quadrilaterals, 93, 94
    investigating angles in triangles, 79, 80
    investigating angles involving parallel lines, 85, 86
    investigating maximum area and minimum perimeter, 57, 58
    investigating rectangles, 47, 48
    investigating scale drawings, 119, 120

graphing calculator,
    determining point of intersection, 245, 246
    drawing curves of best fit, 164–166
    equations and table of values, 217, 218
    multiplying monomials, 273, 274
    solving equations with computer algebra systems (CAS), 225, 226, 228
    subtracting polynomials with CAS, 261, 262
graphs,
    distance-time, 198, 199, 239, 242
    interpreting data in, 181, 182, 183
    scatter plots, 147, 148, 149, 164–166

H
hexagon, 100, 101
hypotenuse, 8

I
integer, 253
interior angles, 76, 86, 88, 89
    in polygons, 99, 100, 101
    in quadrilaterals, 95, 96, 97
    investigating with *The Geometer's Sketchpad*, 79, 93, 103, 104
inverse operations, 8, 227, 228, 229, 232, 234, 237, 285
isosceles triangle, 76, 78
    linear relations in, 172

L
least common multiple, 113
legs of a triangle, 8
like terms, 253, 256
line of best fit, 151, 152, 153, 164, 231
    drawing with *Fathom*, 156–158

linear relations, 169, 170, 172, 198, 212, 222, 223, 233
    determining values in, 231, 232
    first differences in, 191–193, 195
    solving pairs of, 241, 242, 243
    solving problems involving, 235–237, 239

M
maximum area, 53, 54, 55, 66, 128
    investigating with *The Geometer's Sketchpad*, 57
maximum measures, 65, 66
mental math, 228
mind map, 190
minimum measures, 65, 66, 68
minimum perimeter,
    for given area, 61, 62, 63
    investigating with *The Geometer's Sketchpad*, 58
monomials,
    multiplying, 275, 276, 278
    multiplying by a polynomial, 279, 280
    multiplying with CAS, 273, 274

N
negative integers, 253
non-linear relations, 164, 175, 176, 178
    first differences in, 193

O
obtuse triangle, 76
opposite angles, 88

P
parallel lines,
    angles involving, 87, 88, 89, 97
    investigating angles with *The Geometer's Sketchpad*, 85, 86

parallelogram,
    area and perimeter of, 4
partial variation, 211–213, 215, 233
    change in, 221, 222, 223
pentagon, 100
percent (%), 135, 136, 137
perimeter (*see also* minimum perimeter), 3, 4
    of composite figures, 15, 16
    of rectangles, 44
    investigating with *The Geometer's Sketchpad*, 47
point of intersection, 242, 244
polygons,
    angles in, 99, 100
    investigating angles with *The Geometer's Sketchpad*, 103, 104
polynomials,
    adding, 257, 258, 259
    multiplying, 269, 270, 271
    multiplying by a monomial, 279, 280, 282
    subtracting, 263, 264, 265
    subtracting with CAS, 261, 262
positive integers, 253
prisms,
    length of, 24
    volume of, 21, 22, 24, 25, 26
proportional reasoning, 127, 128, 129, 206
proportions, 115, 116, 117
    solving with algebra, 131, 132, 133
Puzzles:
    Grid Paper Pool, 125
    Polygon Pieces, 91
pyramid,
    relating volume to height graphically, 170, 171
    volume of, 25, 26
Pythagorean Theorem, 8, 17, 24, 32

Q

quadratic curve of best fit, 164, 166
quadrilaterals, 100
    angles in, 95, 96
    investigating angles with *The Geometer's Sketchpad*, 93, 94

R

rate (*see also* unit rate), 122
rate of change, 197, 198, 199, 201, 202, 204, 206, 208, 212, 222, 223, 232, 236
ratios, 111, 112, 113
    and percent, 135, 136, 137
    and proportions, 115, 116
rectangles, 53, 54, 55, 66
    area in a composite figure, 12
    area of, 4, 46
    comparing with given areas and perimeters, 49, 50, 51
    investigating area and perimeter with *The Geometer's Sketchpad*, 47, 48
    investigating maximum area and minimum perimeter with *The Geometer's Sketchpad*, 57, 58
    minimum perimeter, 61, 62, 63
    modelling area with algebra tiles, 270, 276
    perimeter, 4, 16, 44,
reflex angle, 102
right triangles, 7, 8, 76
rise, 198, 200, 202, 204, 208, 212
run, 198, 200, 202, 204, 208, 212

S

scale drawings using *The Geometer's Sketchpad*, 119, 120
scatter plots, 147, 148, 149, 152, 153, 164–166
    drawing with *Fathom*, 157
semicircle,
    area of, 13
simplifying expressions, 254, 256, 258, 267, 275
slant height, 31
solutions,
    writing, 146
solving an equation, 227, 228, 229, 283, 284, 285
sphere,
    volume of, 33–35
square,
    area and perimeter of, 4
straight angle, 88, 96
straight line graph, 212

T

terms, 253, 254
TI-83, TI-84, and TI-89 calculators (*see* graphing calculator)
transversals, 85, 86, 88
trapezoid,
    area and perimeter of, 4
    area in a composite figure, 12
trends in graphs, 148, 149, 151, 152, 153, 158, 192
triangle, 100
    angles in, 75, 76, 78, 81, 82, 84
    area and perimeter of, 4
    investigating angles with *The Geometer's Sketchpad*, 79, 80

U

unit rates, 121, 122, 124, 128
unlike terms, 253

V

variables, 253, 276, 283, 284
vertical intercept, 212, 223, 236
volume,
    of a cone, 29, 30
    of prisms and cylinders, 21, 22, 25, 26, 29, 30, 34, 35
    of a pyramid, 25, 26
    of a sphere, 33–35

W

word problems, 110
writing solutions, 146

Z

zero pairs, 253, 254, 263, 264

Answers

Chapter 1 Measuring Figures and Objects

1.1 Measuring Perimeter and Area, page 5

1. a) $P = 28$ m; $A = 45$ m^2
   b) $P \doteq 50.27$ cm; $A \doteq 201.06$ cm^2
   c) $P = 60$ cm; $A = 120$ cm^2
2. a) $P = 26$ cm; $A = 24$ cm^2
   b) $P = 62$ cm; $A = 204$ cm^2
   c) $P \doteq 25.7$ cm; $A \doteq 39.27$ cm^2
3. a) $P = 13$ cm; $A = 10.36$ cm^2
   b) $P = 11.2$ cm; $A = 3.9$ cm^2
   c) $P = 33.3$ cm; $A \doteq 51.8$ cm^2
4. The area of the sail is about 2.6 m^2.
5. About 2356 cm; 23.56 m
   Reanne needs about 24 m of plastic tubing.
6. 369.6 cm^2
7. a) 8 cm   b) 4 cm
8. 6 m

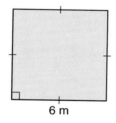

9. a) 120 cm
   b) 105 cm
   The sum of the lengths of the 3 sides is 300 cm.
10. a) About 38 cm
    b) No. Since the diameter of the circle of wire is about 38 cm, it will not fit around a circular tube with diameter 40 cm.

1.2 Measuring Right Triangles, page 9

1. a) 10 cm
   b) 13 cm
   c) About 11.3 m
2. a) About 6.8 cm   b) About 6.5 m
   c) About 16.6 m
3. About 3.4 km
4. $P = 36$ cm

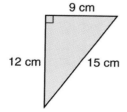

5. a) The offset is about 34 cm long.
   b) The result is reasonable. The offset should be greater than 24 cm.
6. 4.5 m
7. a) Yes; $3^2 + 4^2 = 5^2$
   b) No; $2.5^2 + 5.6^2 \neq 6.4^2$
   c) Yes; $9.6^2 + 12.8^2 = 16^2$
8. a) 12 cm
   b) About 2.5 m
   c) 2 m
9. About 5.7 cm or about 2.8 cm

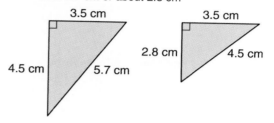

10. a)

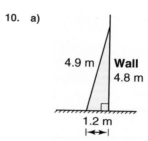

The ladder reaches up the wall about 4.8 m.
Yes. One quarter of 4.8 m is 1.2 m. The ladder is positioned safely.

1.3 Area of a Composite Figure, page 12

1. a) 16 cm^2 b) 12 cm^2
   c) 14 cm^2
2. a) 25 cm^2 b) 48 cm^2
   c) 29 cm^2
3. a) 600 cm^2 b) 24 m^2
   c) 1575 cm^2
4. a) 151 m^2 b) About 317.5 cm^2
5. a) 33 m^2 b) 94.5 m^2
   c) About 31.4 m^2
   The answers are reasonable. Justifications may vary. For example: In part a, the area of the composite figure should be less than the area of a square with side length 6 m. In part b, the area of the composite figure should be a little less than the area of a parallelogram with base 15 m and height 7 m. In part c, the area of the composite figure should be less than the area of the semicircle with radius 6 m, but greater than the area of the semicircle with radius 4 m.
6. a) 13.5 m^2
   b) 135 000 cm^2

ANSWERS 311

c) 900 cm²
d) 150 tiles; 135 000 cm² ÷ 900 cm² = 150

7. a)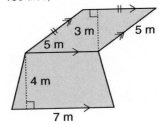
   b) 39 m²
   c) 156 bags
   d) $357.24

1.4 Perimeter of a Composite Figure, page 16

1. a) i) 24 cm
      ii) 24 cm
      iii) 22 cm
      iv) 22 cm
   b) The perimeters in part a are equal in pairs: i and ii, iii and iv.
      All 4 composite figures start from a 6-cm by 4-cm rectangle, but:
      Part i: has an extra equilateral triangle on one of the 4-cm sides
      Part ii: has an equilateral triangle on one of the 4-cm sides removed
      Part iii: has an isosceles triangle on one of the 6-cm sides removed
      Part iv: has an extra isosceles triangle on one of the 6-cm sides
      So, the areas are different.
2. About 26.2 m
3. a) $P \doteq 117.2$ cm
   b) Yes, the result is reasonable. The perimeter of the composite figure should be at least double the perimeter of the interior square.
4. a) About 60.1 cm
   b) About 86.8 cm
5. a) i) About 52.6 m
         40 m of edging are needed for the rectangular patio, about 12.6 m of edging are needed around the circular fish pond.
      ii) About $252
   b) i) About 83.4 m²
      ii) About $3753
   c) I made the assumption that I could buy fractions of a metre of edging and fractions of a square metre of sandstone.

6. Answers may vary. For example:

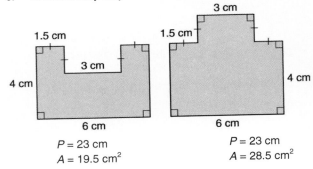

P = 23 cm
A = 19.5 cm²

P = 23 cm
A = 28.5 cm²

7. a) 17 m
   b) About 4 rolls
   c) About $60

Chapter 1 Mid-Chapter Review, page 20

1. a) $A = 6.25$ m²; $P = 10$ m
   b) $A \doteq 6.16$ m²; $P \doteq 8.8$ m
   c) $A = 5.04$ m²; $P = 9.4$ m
2. Answers may vary. For example: Yes, I used the formulas for the area and perimeter of a square, a circle, and a trapezoid.
3. a) $A \doteq 565$ cm²; $P \doteq 30.85$ cm
   b) $A \doteq 15.6$ cm²; $P \doteq 18$ cm
4. a) About 8.1 m   b) About 13.2 cm
5. a) $P \doteq 43$ cm
   b) $P \doteq 43.4$ cm
6. a) $c \doteq 17$ m   b) $h \doteq 6.6$ m
7. a) $P = 52$ cm; $A = 104$ cm²
      Explanations may vary. For example: I know my results are reasonable because the area should be less than the area of a 12-cm by 10-cm rectangle.
   b) $P \doteq 25.4$ m; $A \doteq 40.3$ m²
      Explanations may vary. For example: I know my results are reasonable because the area should be less than the area of a 6-m by 8-m rectangle.
8. Andrew needs just a little more than 6 cans of paint.

1.5 Volumes of a Prism and a Cylinder, page 23

1. a) 216 cm³
   b) 3456 cm³
   c) 512 cm³
2. a) 36 cm³
   b) 192 cm³
   c) 1728 m³
3. a) About 4423.4 m³
   b) About 942.5 cm³
   c) About 1131 cm³

4. a) 486 cm³
    b) Yes. The party-pack box of pasta serves 32 people. If each dimension is doubled, the volume is 2 × 2 × 2, or 8 times as great; so, the party-pack will serve 8 times as many people as the regular size: 4 × 8 = 32
5. a)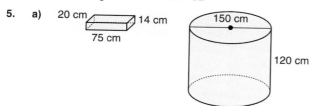
    The cylindrical bale has a greater volume.
    $V_{rectangular\ bale}$ = 21 000 cm³, or 0.021 m³
    $V_{cylindrical\ bale}$ = 2 120 575 cm³, or about 2.1 m³
    b) About 100
6. a) 684 m³
    b) Yes; 0.021 m³ × 1000 = 21 m³
       21 m³ < 684 m³
    c) No; 2.1 m³ × 1000 = 2100 m³
       2100 m³ > 684 m³
7. a) $V \doteq 70$ cm³
    b) $V \doteq 78$ cm³
8. a) About 5132 cm³
    b) About 6.2 kg

1.6 Volume of a Pyramid, page 26

1. Prism: V = 216 cm³; pyramid: V = 72 cm³
2. a) 1500 cm³
    b) About 46.7 m³
    c) 2250 cm³
3. a) 80 m³
    b) 306 cm³
    c) 522.5 m³
4. a)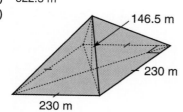
    b) About 167 517 m³ of rock have been lost, perhaps due to wind and water erosion.
5. a) About 11 m³
    b) Answers may vary.
6. a) 208 cm³
    b) Explanations may vary. For example: There is probably some air inside the package as well. Water expands as it freezes.
7. a) About 15.3 cm
    b) About 999.6 cm³
8. About 452 m

1.7 Volume of a Cone, page 31

1. a) 14 cm³
    b) 6.4 cm³
2. a) About 9.42 cm³ (cylinder); about 3.14 cm³ (cone)
    b) About 92.4 cm³ (cylinder); about 30.8 cm³ (cone)
3. a) About 301.6 cm³
    b) About 28.3 cm³
    c) About 194.4 m³
4. About 113.1 cm³
5. a) h = 4 cm; $V \doteq 37.7$ cm³
    b) $h \doteq 7.1$ m; $V \doteq 285.8$ m³
    c) $h \doteq 7.5$ cm; $V \doteq 502.7$ cm³
6. a) About 13.2 m³
    b) Answers may vary. For example: It should be one-third the volume of a related cylinder.
7. About 153 million cubic metres, or 0.15 km³

1.8 Volume of a Sphere, page 35

1. Cylinder: about 785.4 cm³; sphere: about 523.6 cm³
2. About 1436.8 cm³
3. a) About 4188.8 cm³
    b) About 268.1 m³
    c) About 179.6 cm³
4. a) About 3053.6 m³
    b) About 1436.8 cm³
    c) About 310.3 cm³
5. a) About 4 breaths; I assumed the student emptied only part of his lungs with one breath.
    b) Answers will vary. For example: It usually takes me 5 or 6 breaths to blow up a balloon.
6. a) 2744 cm³
    b) 14 cm
    c) About 1436.8 cm³
    d) 1307.2 cm³
    Answers may vary. For example: I assumed Lyn will make the largest possible wooden ball and that her ball will be spherical.
7. a) About 19 scoops
    b) About 23¢ per scoop
    c) 17¢ per cone; single-scoop ice-cream cone: 40¢ (or about 50¢); double-scoop ice-cream cone: 63¢ (or about 75¢). I rounded up to the nearest quarter so Meighan could make a profit.

Chapter 1 Review, page 38

1. a) P = 30 cm; A = 40 cm²
    b) $C \doteq 37.7$ cm; $A \doteq 113.1$ cm²
2. a) 4356 cm²
    b) 3744 cm²
3. a) 15 cm
    b) 18 cm
4. About 4.9 m

5. About 61.6 cm²
6. About 5731.83 cm²
7. About 81.2 cm
8. a) About 51.4 cm
   b) The answer is reasonable. The perimeter should be a little longer than the perimeter of a 10-cm by 10-cm square; that is, a little larger than 40 cm
9. a) 1820 cm³
   b) About 2152.8 cm³
10. a) 16 people; the new volume is 2 × 2, or 4 times as great. 4 people × 4 = 16 people
    b) 32 people; the new volume is 2 × 2 × 2, or 8 times as great. 4 people × 8 = 32 people

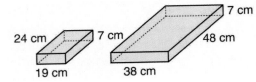

11. 560 cm³
12. a) 1880 m³
    b) The volume of the pyramid is one-third the volume of a square prism with a 20-m by 20-m square base and height 14.1 m.
    $V_{pyramid} = \frac{1}{3} \times V_{prism}$
    $1880 = \frac{1}{3} \times 5640$
    The answer is reasonable.
13. a) 28 cm³
    b) 6 cm
    c) 2 cm; 14 cm² × 2 cm = 28 cm³
14. $h \doteq 51.5$ cm; $V \doteq 26\ 102.4$ cm³
15. a) About 2482.7 cm³
    b) About 7238.2 m³
16. a) About 43 396.8 cm³
    b) Answers may vary. For example: the volume of the sphere is two-thirds the volume of a cylinder into which the sphere just fits.

Chapter 1 Practice Test, page 40

1. C
2. C
3. a) About 16.6 cm
   b) About 8.9 m
4. a) The hypotenuse of the right triangle: about 640 m; park's area: 292 000 m²
   b) About 2151 m
5. No. $24^2 + 17^2 \neq 29^2$
6. About 12 bags. I assumed that there was no room left between the bags in the cylindrical cooler.

Chapter 2 Investigating Perimeter and Area of Rectangles

2.1 Varying and Fixed Measures, page 45

1. a) The lengths of line segments AP and PB vary.
   b) The length of line segment AB stays the same.
2. a) The measures of arc AB and of ∠BOA vary.
   b) The lengths of radii OA and OB stay the same. The circumference stays the same.
3. a) The lengths of line segments AC, AD, and AB, and the measures of ∠BAC, ∠DAB, and ∠CBA vary.
   b) The lengths of line segments BC and CD and the measure of ∠ACB remain the same.
4. a) $P = 120$ cm, $A = 875$ cm²
   b) Answers may vary. For example:
      The perimeter will stay the same (the sum of width and length stays the same). The area will change.

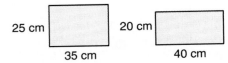

   c) $P = 120$ cm, $A = 800$ cm²
      Yes. The perimeter stays the same.
      The area decreases.
5. a) The perimeter will increase. The area will stay the same, since length × width stays the same.
   b) Yes. $P = 165$ cm, $A = 875$ cm²

6. a) Answers may vary. For example:
      1 unit by 9 units, 2 units by 8 units, 3 units by 7 units, 4 units by 6 units, 5 units by 5 units

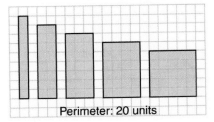

   b) Answers may vary. For example:
      1 unit by 15 units, 2 units by 14 units, 3 units by 13 units, 4 units by 12 units, 5 units 11 units

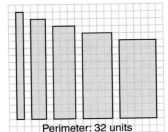

7. a) Answers may vary.

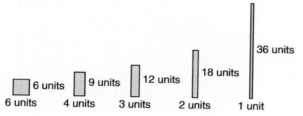

b) Answers may vary.

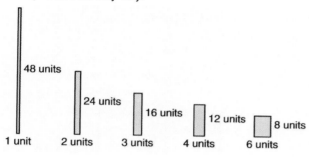

8. a)

Length (cm)	Width (cm)	Perimeter (cm)	Area (cm^2)
4	4	16	16
5	3	16	15
6	2	16	12
7	1	16	7

b) No. Rectangles with the same perimeter have different areas.

c)

Width (cm)	Length (cm)	Perimeter (cm)	Area (cm^2)
4	4	16	16
2	8	20	16
1	16	34	16

d) No. Rectangles with the same area have different perimeters.

9. a) The area stays the same if one dimension is multiplied by a number and the other dimension is divided by the same number. For example: the length is doubled and the width is halved.

b) I list factors whose product equals the given area.

10. a) The lengths of line segments AB and BC, and the measures of ∠A and ∠B change.

b) Line segment AC and ∠C stay the same.

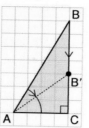

2.2 Rectangles with Given Perimeter or Area, page 51

1. a)

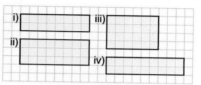

b) i and iii have the same perimeter, 20 cm. The sum of the length and width is equal, 10 cm.
ii and iv have the same perimeter, 22 cm. The sum of the length and width is equal, 11 cm.

c) ii and iii have the same area. The product of the length and width is 24 cm^2.

d) Yes. For example: i and iii have the same perimeter, 20 cm, but different areas: i) 16 cm^2 and iii) 24 cm^2

e) Yes. For example: ii and iii have the same area, 24 cm^2, but different perimeters: ii) 22 cm and iii) 20 cm

2. a), b)

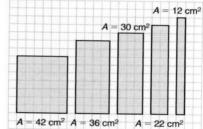

d) As area increases, the dimensions get closer in value, and the rectangles become more like a square.

3. a), b)

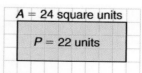

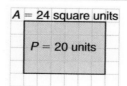

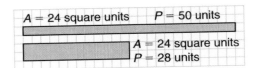

d) As the perimeter increases, the difference between length and width increases, and the rectangles become long and narrow.

4. a)
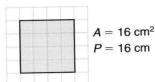

b)
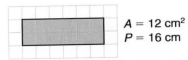

c) The square has the greatest area.

5. a) Each rectangle has area 36 cm².
   b) Rectangle C has the least perimeter. Its length and width are closest in value. Rectangles A and E have the greatest perimeter. The difference between their length and width is the greatest.
   c) A: 30 cm; B: 26 cm; C: 24 cm; D: 26 cm; E: 30 cm
   Yes. Rectangle C has the least perimeter, 24 cm. Rectangles A and E have the greatest perimeter, 30 cm.

6. a) Each rectangle has perimeter 18 cm.
   b) Rectangles D and E have the greatest area. Their length and width are closest in value. Rectangles A and H have the least area. The difference between their length and width is the greatest.
   c) A: 8 cm²; B: 14 cm²; C: 18 cm²; D: 20 cm²; E: 20 cm²; F: 18 cm²; G: 14 cm²; H: 8 cm²
   Yes. Rectangles D and E have the greatest area, 20 cm². Rectangles A and H have the least area, 8 cm².

7. No. Two rectangles with the same perimeter and the same area are congruent.

2.3 Maximum Area for a Given Perimeter, page 55

1. a), b)
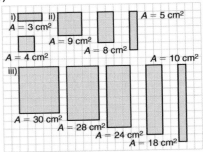

   c) i) 2 cm by 2 cm
      ii) 3 cm by 3 cm
      iii) 6 cm by 5 cm

2. a), b)
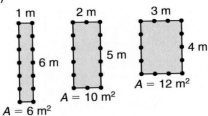

   c) The 3-m by 4-m garden has the maximum area. Its length and width are closest in value. The garden cannot be a square. The sides are whole numbers only.

3. a) 81 cm² (9 cm by 9 cm)
   b) 225 cm² (15 cm by 15 cm)
   c) 351.5625 cm² (18.75 cm by 18.75 cm)

4. The maximum area occurs when the rectangle is a square, with dimensions 10.5 m by 10.5 m. The greatest area for a rectangle with perimeter 42 m is 110.25 m².

5. Explanations may vary.
   For example: Yes. For a given perimeter, the maximum area occurs when length and width are closest in value (the rectangle is as close to a square as possible). Among all possible rectangles with perimeter 16 cm, the 4-cm by 4-cm square has the maximum area.

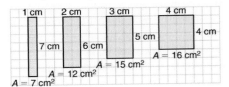

316 ANSWERS

6. a) A table that seats 20 people can be represented as a rectangle with perimeter 20 units. That is, the sum of its length and width is 10. There are 5 possible rectangular arrangements: 1 by 9, 2 by 8, 3 by 7, 4 by 6, 5 by 5.

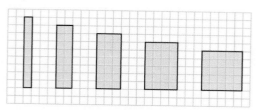

b) The arrangement with greatest area (5 by 5) requires the most tables: 25. The arrangement with least area (1 by 9) requires the fewest tables: 9

c) If you use the 1 by 9 table, the person at one end of the table won't be able to speak to the person at the other end. The square is closest to a circle, which is the best shape for communicating around a table.

7. a) Area of square flowerbed: 4 m²; area of circular flowerbed: about 5.1 m²

b) Bonnie's; the area of a circle with perimeter P is always greater than the area of a rectangle with the same perimeter. Students' answers should include drawings.

Chapter 2 Mid-Chapter Review, page 60

1. a) Both the area and the perimeter increase.
   b) Both the area and the perimeter decrease.
   c) Both the area and the perimeter increase.
   d) Both the area and the perimeter decrease.
   e) The perimeter stays the same and the area increases.

2. a) All rectangles have the same perimeter: 16 cm.
   b) The area varies. From least to greatest: 7 cm², 12 cm², 15 cm², 16 cm². The 4-cm by 4-cm square has the greatest area.

3. a) i) 1 m by 30 m, 2 m by 15 m, 3 m by 10 m, 5 m by 6 m
   ii) 1 m by 64 m, 2 m by 32 m, 4 m by 16 m, 8 m by 8 m
   b) i) 62 m, 34 m, 26 m, 22 m
   ii) 130 m, 68 m, 40 m, 32 m

4. Students' answers should include diagrams and calculations.
   a) The perimeter stays the same. The area decreases by 27 cm².
   b) The area stays the same. The perimeter doubles.

5.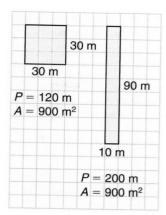

a) $A_{\text{first paddock}}$ = 30 m × 30 m = 900 m²
   $A_{\text{second paddock}}$ = 10 m × 90 m = 900 m²

b) No; 120-m of fencing are needed for the 30-m by 30-m paddock. 200-m of fencing are needed for the 10-m by 90-m paddock.

6.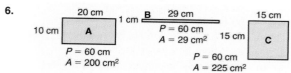

7. The maximum area is 2025 m² (for a 45-m by 45-m square lot).

2.4 Minimum Perimeter for a Given Area, page 63

1. a)

   6 cm

   A

   10 cm $P = 32$ cm
   $A = 60$ cm²

   b)

   5 cm

   $P = 34$ cm   12 cm
   $A = 60$ cm²

c) No. The perimeter of the 5-cm by 12-cm rectangle is greater than the perimeter of rectangle A.

2. a)

   B
   $P = 52$ cm   8 cm
   $A = 144$ cm²

   18 cm

b)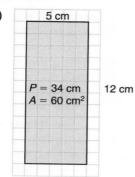

16 cm

$P = 50$ cm   9 cm
$A = 144$ cm²

c) The 12-cm by 12-cm square has the minimum perimeter for the given area.

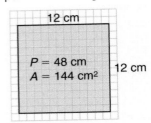

3. a) The minimum perimeter occurs when length and width are closest in value.
   i) 5 by 5, P = 20 units
   ii) 10 by 5, P = 30 units
   iii) 15 by 5, P = 40 units
   iv) 10 by 10, P = 40 units
   b) i) P = 600 cm
   ii) P = 900 cm
   iii) P = 1200 cm
   iv) P = 1200 cm

4. a) Start at 25. The area increases by 25 stones each time.
   b) No. There is no pattern in the perimeters.

5. a) About 21.9 m
   b) About 31.0 m
   c) About 38.0 m
   d) About 43.8 m

6. Both the length and width of a rectangle with the least possible perimeter are about 31.6 m. The least possible perimeter for a rectangle with area 1000 m² is approximately 126.5 m.

7. a) The minimum perimeter occurs when the length and width are closest in value: 8 tiles by 7 tiles. The minimum perimeter is 30 units.
   b) No; 56 tiles do not form a square. We cannot use fractions of a square tile.

8. No. Explanations may vary. For example: For a given area, rectangles whose lengths and widths are closest in value have the least perimeter.

9. a)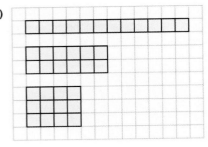

b) The arrangement that seats the most people is the rectangle with the maximum perimeter: 12 by 1 arrangement
The arrangement that seats the fewest people is the rectangle with the minimum perimeter: 4 by 3 arrangement

c) The rectangular arrangement with the minimum perimeter seats the fewest people.

10. a), b) Answers may vary.

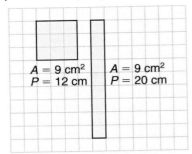

c) No. For a given area, a square has the minimum perimeter.

2.5 Problems Involving Maximum or Minimum Measures, page 67

1. a) 8 m; 6 m; 4 m; 2 m; students' answers should include diagrams.
   b) 4 m by 8 m; A = 32 m²; the length is twice the width.

2. a) Students' answers should include diagrams of 9 possible rectangles: 1 m by 18 m; 2 m by 16 m; 3 m by 14 m; 4 m by 12 m; 5 m by 10 m; 6 m by 8 m; 7 m by 6 m; 8 m by 4 m; 9 m by 2 m.

b)

Width (m)	Length (m)	Area (m²)
1	18	18
2	16	32
3	14	42
4	12	48
5	10	50
6	8	48
7	6	42
8	4	32
9	2	18

c) 5 m by 10 m; the length is twice the width.

3. a) i) 20 m by 20 m

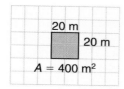

ii) 20 m by 40 m

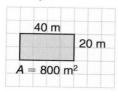

b) i) Length equals width.
ii) Length is twice the width.

4. The 6-units by 12-units arrangement has the minimum distance: 24 units.

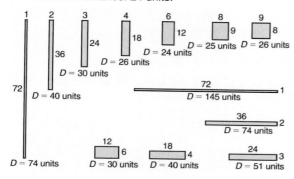

5. The minimum edging is required when the dimensions are closest in value: 8 units by 9 units.

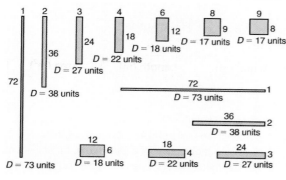

6. a)

Sections along each of 3 sides	Sections along each of 2 sides	Overall area (m²)
2	10	20
4	7	28
6	4	24

b) 4 sections along 3 sides, and 7 sections along the other 2 sides for a maximum area of 28 m².

Chapter 2 Review, page 70

1. a) The length of the ladder stays the same. The right angle between the floor and the wall remains the same.
   b) The distance from the top of the ladder to the floor decreases; the horizontal distance from the ladder to the wall increases. The angle that the ladder makes with the floor decreases, the angle that the ladder makes with the wall increases.

2. a)

Width (cm)	Length (cm)	Area (cm²)
5	45	225
10	40	400
15	35	525
20	30	600
25	25	625

b) For a rectangle with perimeter 100 cm, with each 5-cm increase of the width, the length decreases by 5 cm. The sum of the width and length is always equal to 50 cm. The area increases as the length and width become closer in value.

3. a)

Width (cm)	Length (cm)	Perimeter (cm)
2	36	76
4	18	44
6	12	36
8	9	34

b) If the width is multiplied by a number, the length is divided by the same number. The perimeter decreases as the length and width become closer in value.

4. a) $P = 16$ cm; $A = 16$ cm²
   b) There are 3 possible rectangles with the same perimeter as square B: 1 cm by 7 cm, 2 cm by 6 cm, 3 cm by 5 cm

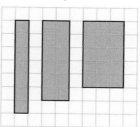

   c) No. For a given perimeter, the maximum area occurs when the rectangle has equal length and width (it is a square).

5. a) OABC and ODEF have equal perimeters: 24 units, but different areas.
   A_{OABC} = 32 square units,
   A_{ODEF} = 27 square units
   b)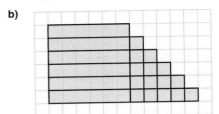
   c) For a given perimeter, when the length decreases by 1 unit, the width increases by 1 unit.
   d) i) A_{max} = 36 square units (6 units by 6 units)
      ii) A_{min} = 11 square units (1 unit by 11 units)
   e) No for part d) i): For a given perimeter of 24 units, the rectangle with the maximum area is always the one that has equal length and width (the 6-units by 6-units square).
      Yes for part d) ii): More rectangles are possible if the width can be less than 1 unit.
      For example: 0.5 units by 11.5 units
6. a) i) 2 cm by 2 cm
      ii) 8.5 cm by 8.5 cm
      iii) 13.5 cm by 13.5 cm
   b) For a rectangle with a given perimeter, the maximum area occurs when the dimensions are equal.
      i) 4 cm^2
      ii) 72.25 cm^2
      iii) 182.25 cm^2
7. a) 14.5 m by 14.5 m
   b) 210.25 m^2
8. a) 14 m by 15 m
   b) 210 m^2
9. a) i) 4 m by 4 m
      ii) About 4.9 m by 4.9 m
      iii) About 7.1 m by 7.1 m
   b) For a rectangle with a given area, the minimum perimeter occurs when the dimensions are closest in value.
      i) 16 m
      ii) 19.6 m
      iii) 28.4 m
10. 14 cm by 14 cm; P = 56 cm

11. a)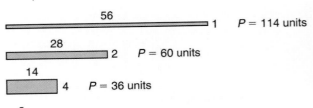
    b) The 8-m by 7-m arrangement requires minimum fencing. The dimensions are closest in value.
12. a) 50 m
    b) 230 m
    c) 100 m by 200 m; the maximum area is 20 000 m^2. Given a fixed length for 3 sides of a rectangle, the rectangle with greatest area has length double its width.

Chapter 2 Practice Test, page 72

1. D
2. A
3. a)
   1 ▭ 24 P = 50 units
   2 ▭ 12 P = 28 units
   3 ▭ 8 P = 22 units
   4 ▭ 6 P = 20 units
   b) The patio with dimensions that are closest in value requires the minimum amount of edging: 4 m by 6 m (P = 20 units).
4. a) 49 cm^2
   b) 462.25 cm^2
5. Students' answers should include diagrams.
   a) For a rectangle with a given area, the minimum perimeter occurs when the dimensions are closest in value.
   b) For a rectangle with a given perimeter, the maximum area occurs when the dimensions are closest in value.
6. a) 4 sections by 5 sections. The maximum area is 20 m^2.
   b) Workers should use the 2 sections to make a 5-m by 5-m rectangular storage, with an area of 25 m^2. The maximum area occurs when the rectangle has equal length and width (it is a square).
7. 4 m by 10 m, or 5 m by 8 m

Chapter 3 Relationships in Geometry

3.1 Angles in Triangles, page 77

1. This shows the property that the sum of the angles in a triangle is 180°.
2. a) ∠C = 50°
   b) ∠E = 46°
   c) ∠K = 23°
3. a) Yes, since 41° + 67° + 72° = 180°
   b) No, since 40° + 60° + 100° ≠ 180°
   c) No, since 100° + 45° + 30° ≠ 180°
4. No, a triangle cannot have two 90° angles. Since 90° + 90° = 180°, the third angle would be 0°. Also, if there are two 90° angles then two sides would be parallel, and they can never intersect, so the triangle cannot be formed.
5. a) The acute angles in a right triangle have a sum of 90°.
   b) The sum of the angles in any triangle is 180°. If one of the angles is a right angle, the other two angles must have a sum of 180° − 90° = 90°.
   c) i) ∠D = 30°
      ii) ∠P = 36°
      iii) ∠C = 45°
6. The angles opposite the equal sides in an isosceles triangle are equal.
7. a) i) ∠A + 50° + ∠B = 180°
      ii) ∠A = ∠B = 65°
   b) i) ∠A + ∠B + 90° = 180°
      ii) ∠A = ∠B = 45°
   c) i) ∠A + ∠B + 31° = 180°
      ii) ∠A = 118°, ∠B = 31°
   d) i) ∠A + ∠B + 66° = 180°
      ii) ∠A = 48°, ∠B = 66°
8. a) 100°
   b) Since each of the lower angles increases by 2° from 40° to 42°, the upper angle decreases by 4° from 100° to 96°. The upper angle is 96°.
9. 180° ÷ 3 = 60°; each angle measures 60°.
10. a) ∠A = 97°   ∠B = 83°   ∠C = 69°
    b) ∠D = 62°   ∠E = 36°
    c) ∠F = 80°   ∠G = 100°   ∠H = 40°

3.2 Exterior Angles of a Triangle, page 83

1. i) a) ∠A and ∠B
      b) ∠BCD = ∠A + ∠B
   ii) a) ∠D and ∠E
       b) ∠DFG = ∠D + ∠E
   iii) a) ∠H and ∠K
        b) ∠KJM = ∠K + ∠H
2. a) ∠CAD = ∠B + ∠C
      = 118°
   b) ∠EGH = ∠E + ∠F
      = 111°
   c) ∠JKM = ∠J + ∠H
      = 72°
   d) ∠MRP = ∠M + ∠N
      = 92°
   e) ∠QST = ∠Q + ∠R
      = 143°
   f) ∠XVU = ∠X + ∠W
      = 120°
3. a) ∠BAC = 55°, ∠C = 55°
   b) ∠F = 79°, ∠FGE = 60°
   c) ∠H = 49°, ∠JKH = 109°
4. The plate angle is 40°; each exterior angle of a triangle is equal to the sum of the two opposite interior angles.
5. ∠HJK = 28°, ∠GHK = 56°, ∠DEK = 90°, ∠DFE = 72°, ∠ABF = ∠JHK = 124°, ∠BFA = ∠HKJ = 28°, ∠BCF = ∠HGK = ∠BFC = ∠HKG = 62°, ∠FCD = ∠KGD = 118°, ∠CDF = ∠GDK = 44°, ∠CFD = ∠GKD = 18°, ∠FDE = ∠KDE = 18°
6. a) 128°

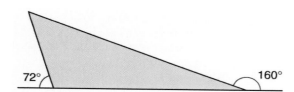

   b) The sum of the 3 exterior angles is 360°.
   c) The triangles have the same angles but the triangles have different sizes.
7. a) x = 125°
   b) x = y = 145°
   c) x = 47°, y = 133°, z = 142°
8.

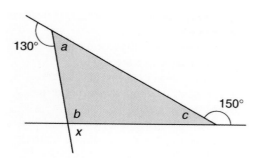

For example, 130° + 150° + x = 360°; x = 80°
a + 130° = 180°; a = 50°
c + 150° = 180°; c = 30°
b = 180° − a − c = 100°
Also, a + b = 150°; b + c = 130°, c + a = x
Or, a + b + c = 180°

3.3 Angles Involving Parallel Lines, page 89

1. I used these relationships:
   • opposite angles are equal

- angles that form a straight angle have a sum of 180°
- when parallel lines are intersected by a transversal, corresponding angles are equal
  a) ∠BCH = ∠GCD = ∠EBC = 62°,
  ∠ABE = ∠FBC = ∠BCG = ∠DCH = 118°
  b) ∠EBC = 62°
  ∠ABE = ∠FBC = 118°
2. a) b = 60°, a = 15°
   b) d = e = 86°, f = 94°, g = 141°, h = 39°
   c) a = 55°, b = c = 65°, d = 120°
3. Yes, Petra is correct.

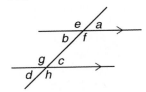

- a = b = c = d
  So, all the acute angles are equal.
- e = f = g = h
  So, all the obtuse angles are equal.
- c + f = b + g = 180°
  e + b = a + f = 180°
  So, the sum of an acute angle and an obtuse angle is 180°.
4. ∠BCD = 133°
5. 34° + 145° = 179°, not 180°, so the boards are not parallel.
6. ∠CED = 88°, ∠ECD = ∠EDC = 46°
7. a) The sum of the interior angles is 180°.
      The sum of the exterior angle and the adjacent interior angle is 180°.
      The exterior angle is equal to the sum of the opposite interior angles.
      The sum of the exterior angles of a triangle is 360°.
   b) Corresponding angles are equal.
      Alternate angles are equal.
      Interior angles have a sum of 180°.
   c) Students' answers should include a design.
8. a) a + c = b
   b) Yes. As long as the top and bottom lines are parallel, a, b, and c are always related so that a + c = b.
   c) If you draw a third parallel line through B, and use the property that alternate angles are equal, the sum of a and c is always equal to b.

Chapter 3 Mid-Chapter Review, page 92

1. When you arrange the three angles of a triangle next to each other, they form a straight line. That is, the angles in a triangle have a sum of 180°.

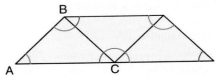

2. a = 50°, b = 40°
3. Students' answers may show diagrams with congruent angles labelled.
4. a) a = c = 58°, b = 122°
   b) e = 115°, f = 65°, g = 72°, h = 43°
5. Angle measures will vary according to students' drawings.
6. x = 55°, y = 125°, z = 55°
   Assumptions: The flag is a rectangle; the horizontal line segments are parallel.
7. a) a = b = c, c = e, f + b + e = 180°, g + h = 90°,
      e + d = 90°, a = f, c + d = 90°, g + 90° = 143°
   b) a = b = c = 60°, d = 30°, e = 60°,
      f = 60°, g = 53°, h = 37°

3.4 Interior and Exterior Angles of Quadrilaterals, page 97

1.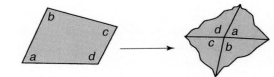

2. a) a = 75°   b) b = 27°   c) c = 61°
3. a) They have a sum of 180°.
   b) The angles that are between the parallel lines on the same side of a transversal are interior angles and have a sum of 180°.
4. a) a = 91°   b = 89°   c = 91°
   b) d = 53°   e = 127°   f = 53°
   c) h = 132°   j = 48°   k = 132°
5. The opposite angles in a parallelogram are equal.
6. a), b) The sum of all the angles is 360°, because it is a quadrilateral. If we draw the diagonal that forms 2 congruent triangles, we see that one pair of opposite angles are equal.
7. a) a = 86°
   b) b = 54°
   c) d = 124°
8. a) ∠R + ∠O + ∠S + ∠E = 360°
      ∠R = ∠E, ∠O = ∠S
      ∠O + ∠R = 180°, ∠S + ∠E = 180°
   b) Students can measure angles with a protractor to demonstrate the relationships in part a.

3.5 Interior and Exterior Angles of Polygons, page 101

1.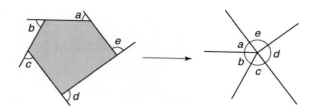

2. **a)** 900° **b)** 1440° **c)** 3960°
3. **a)** $x = 148°$ **b)** $x = 55°$
4. **a)** 108° **b)** 135°
   c) 144° **d)** 162°
5. **a)** 72° **b)** 45° **c)** 36° **d)** 18°
6. **a)** The Canadian loonie has 11 sides.
   b) About 147°
7. **a)** The sum S of the interior angles in an n-sided polygon is $S = (n − 2) × 180°$.
   If the polygon is a regular polygon, then the measure of each interior angle is:
   $$\frac{(n-2) \times 180°}{n}$$
   The exterior angles in any polygon have a sum of 360°.
   If the polygon is a regular polygon with n sides, then the measure of each exterior angle is:
   $$\frac{360°}{n}$$
   At each vertex, the interior angle and the exterior angle have a sum of 180°.
   b) Students' answers should include designs.
8. **a), b)** Students' answers should include drawings.
   c) Quadrilateral: 360°, pentagon: 540°
   d) Yes, the number of triangles the polygon can be divided into is still 2 less than the number of sides.
9. **a)** Regular pentagons and regular hexagons
   b) 120°
   c) 108°
   d) 348°; the sum is close to, but not equal to, 360°. This is because the surface is curved.
10. **a)** $6 × 180° = 1080°$
    b) 360°; this is the sum of the angles at the point inside the hexagon. These angles are *not* interior angles of the hexagon.
    c) $n × 180° − 360° = (n − 2) × 180°$

Chapter 3 Review, page 106

1. If you cut off the corners (angles) of a triangle and arrange them side by side, they form a straight angle, which has a measure of 180°.
2. No. If a triangle had 2 obtuse angles, the sum of those two angles would be greater than 180°.
3. **a)** $x = 59°$
   b) $x = y = 69°$
   c) $x = 48°$
4. $a = 44°, b = 44°, c = 63°, d = 63°, e = 63°$
5. **a)** No. Alternate angles are not equal.
   b) Yes. The interior angles 116° and 64° have a sum of 180°.
6. $x = 49°$ $y = 49°$
7. **a)** $x = 89°$
   b) $x = 21°, y = 40°, z = 119°$
   c) $y = 85°$
   d) $x = 105°$
8. **a)** 1080°
   b) i) 135°
      ii) 45°
9. **a)** 720°
   b) 1800°
   c) 2880°
10. **a)** 540°
    b) The sum of the exterior angles of a pentagon is 360°.
    c) Yes
11. Interior angle: 176.4°; exterior angle: 3.6°

Chapter 3 Practice Test, page 108

1. B
2. B
3. Cut out and arrange the corners next to each other; draw a diagonal and add the interior angles of the 2 triangles formed by the diagonal.
4. **a)** $a = 81°$
   b) $a = 34°, b = 49°, c = 34°, d = 131°$
   c) $x = 135°, y = 45°$
5. **a)** 540°
   b) 135°
6. ∠PRS = ∠PSR (because it is an isosceles triangle)
   ∠PRQ + ∠PRS = 180°
   ∠RST + ∠PSR = 180°
   So, ∠PRQ = ∠RST

Chapter 4 Proportional Reasoning

4.1 Equivalent Ratios, page 112

1. **a)** 4:6 **b)** 8:6 **c)** 12:15 **d)** 30:25
2. **a)** 3:5 **b)** 3:1 **c)** 5:4 **d)** 1:2
3. **a)** 4:10, 20:50 **b)** 1:4, 25:100
   c) 20:2, 100:10 **d)** 150:100, 15:10
4. **a)** 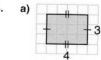 **b)**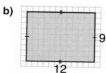

   c) No, 18:26 is not equivalent to 6:8.
5. The ramp should have dimensions in the ratio 2:24. It could be 20 cm high and 240 cm long.

6. a) 24:16, 24:15
    b) Megan's punch uses less ginger ale for the same amount of concentrate.
7. a) Punch B; it uses concentrate and water in the ratio 5:8, punch A uses a ratio of 4:8.
    b) Punch B; it uses concentrate and water in the ratio 5:6, punch A uses a ratio of 4:6.
8. a) Set B; explanations may vary.
    b) Set B;
       Set A uses blue and clear liquid in the ratio 3:2.
       Set B uses blue and clear liquid in the ratio 4:2.
    c) Sketches may vary. The ratio of blue to clear liquid must be equivalent to 3:2. For example:

9. 15 cm by 24 cm
10. a) 0.5 cm   b) 0.4 cm

4.2 Ratio and Proportion, page 117

1. a) First terms: $125 = 25 \times 5$; first ratio: $20 = 4 \times 5$
    b) First terms: $120 = 12 \times 10$; first ratio: $10 = 10 \times 1$
    c) Second terms: $100 = 4 \times 25$; first ratio: $75 = 3 \times 25$
    d) Second terms: $48 = 16 \times 3$; first ratio: $3 = 3 \times 1$
2. a) $n = 20$ b) $n = 3$ c) $n = 3$ d) $n = 30$
3. a) $z = 6$ b) $z = 15$ c) $z = 7$ d) $z = 8$
4. Yes, the ratios are equivalent, so the number of the songs is proportional to the amount of memory.
5. 18 L
6. 25 h
7. 15 potatoes
8. a) $c = 45$ b) $n = 30$
    c) $y = 8$ d) $z = 15$
9. a) 84 cm
    b) Calculate $28 \div 17 \doteq 1.65$ and then $51 \times 1.65 \doteq 84$.
10. 276 times
11. a) 84 teeth   b) 189 teeth
12. a) 500 girls, 400 boys b) 15 girls, 12 boys

4.3 Unit Rates, page 123

1. a) 2 goals scored per game
    b) $10 per hour   c) $0.50 per orange
    d) 110 km/h
2. a) $0.50 per CD   b) $0.79 per apple
    c) 1.5 kg lost per week
    d) 4.4 km/h
3. 150 L/h
4. The hardwood is more expensive at $44.12/m².
5. e-Tunes is the most economical music club at $1.05 per song.
6. No. The can from the machine costs 3 times as much as the cans in the 12-pack.
7. Can C

8. a) 144 tea bags for $5.39
    b) Use equivalent ratios.
    c) Sue might not use 144 bags of tea before they go stale.
9. a) Price (in dollars) per 100 g
    b) $0.62 per 100 g, $0.57 per 100 g
    c) Dee's Delight

Chapter 4 Mid-Chapter Review, page 126

1. a) equivalent   b) not equivalent
    c) not equivalent   d) equivalent
2. a) Pitcher B
    b) A pitcher containing 6 parts concentrate and 4 parts water
3. No, the ratios are not equivalent.
4. a) $n = 40$ b) $m = 20$
    c) $y = 3$ d) $r = 12$
5. 160 mL
6. a) 3:10
    b) 168 male doctors
    c) 51 female doctors
    d) The ratio of female doctors to male doctors remains the same.
7. a) 15 km/L   b) 30 words/min
    c) $18.50/h   d) 87.5 km/h
8. Beckie
9. a) $1.04, $1.08   b) Cereal A

4.4 Applying Proportional Reasoning, page 129

1. a) $15   b) $195
2. a) $56   b) $616
3. $480
4. a) 14 cases   b) Use equivalent ratios.
5. a) 300 km   b) 10 L
6. a) 7500 books b) 58 min
7. a) 750 cm, or 7.5 m b) 100 cm
8. a) 45 points   b) 38 points
    c) Answers may vary. For example: Chloë and her father scored 19 baskets each.
9. a) $191.25 US
    b) $11.76 Can; The exchange rate remains the same.
10. Too dry

4.5 Using Algebra to Solve a Proportion, page 133

1. a) $n = 9$ b) $n = 3$ c) $n = 15$
    d) $n = 40$ e) $n = 10$
2. a) $c = 27$ b) $m = 7$
    c) $y = 105$ d) $a = 175$
3. a) 0.83 m   b) 7.4 m
4. $445.00
5. a) 160 000 tickets b) 23.44 min
6. a) 2.03 m   b) 95.4 cm
    c) Answers may vary.

4.6 Percent as a Ratio, page 137

1. a) 0.07 b) 0.15 c) 0.35
   d) 0.8 e) 1.2
2. a) $36.50 b) 12.5 kg
   c) 14 m d) 150 g
3. $111.99
4. 10 000 vehicles
5. a) $8.40 b) $68.39
6. $219.45
7. a) 216 spectators b) 173 spectators
8. $300
9. a) $6.67 b) $506.67
10. a) $53.26 b) $953.26
11. 8% per year

Chapter 4 Review, page 140

1. a) 6:10, 9:15 b) 6:7, 12:14
   c) 3:2, 6:4 d) 45:7, 90:14
2. a) No
   b) Any spinner with the same ratio of red sectors to total sectors would be equivalent. For example, a spinner with 16 sectors, 6 of which are red or 24 sectors, 9 of which are red
3. Mix A. It uses more concentrate for the same amount of water.
4. a) $n = 12$ b) $a = 7$
   c) $e = 42$ d) $m = 24$
5. 240 mL
6. a) $1\frac{1}{2}$ cups b) 45 cookies
7. a) 48 km/h b) 35 words/min
   c) $0.40/min d) $1.58/kg
   e) 30 pages/min
8. B: 130 mL for $1.99 is the better buy.
9. a) 9 kg b) $42.40
10. a) About 9 days b) 22.5 rows
11. a) 360 km b) 35 L
12. 15 copies
13. a) $11.40 b) $53.35
    c) $27.00 d) $9.47
14. a) $51.99 b) $59.27
15. 84 h
16. a) $8.75 b) $358.75
17. a) $90 b) $1590

Chapter 4 Practice Test, page 142

1. C
2. C
3. a) $a = 25$ b) $b = 5$
   c) $c = 20$ d) $n = 8.4$
4. 18 oranges
5. a) $12.50 b) $2512.50
6. You can measure your height and the length of your shadow. Then measure the length of the shadow of the flagpole and use equivalent ratios to find the height of the flagpole.

Cumulative Review Chapters 1–4, page 143

1. a) $P = 30$ cm, $A = 30$ cm^2
   b) $P = 24$ cm, $A = 28$ cm^2
2. $A \doteq 137.1$ cm^2
   $P \doteq 52.6$ cm
3. a) 108 cm^3 b) About 22 m^3
   c) About 205.3 cm^3 d) About 904.8 cm^3
4. a) 112 cm^3 b) 12 cm c) 4 cm
5. 1 cm, 20 cm; 2 cm, 10 cm;
   4 cm, 5 cm; 8 cm, 2.5 cm
6. a) 4.5 m by 4.5 m b) 20.25 m^2
   c) i) 4.5 m by 9 m ii) 40.5 m^2
7. a) i) 5 cm by 5 cm
      ii) About 5.7 cm by 5.7 cm
      iii) About 8.8 cm by 8.8 cm
   b) i) 20 cm ii) About 22.8 cm
      iii) About 35.2 cm
   For a given area, the rectangle with the minimum perimeter is always a square.
8. 73° and 73°, or 34° and 112°
9. a) $q = 77°$, $p = 135°$, $r = 58°$, $s = 122°$
   b) $a = c = y = 30°$,
      $b = d = x = z = 150°$
   c) $a = b = 105°$, $c = d = 75°$
10. a) 720°
    b) The hexagon can be divided into 4 triangles. Each triangle has interior angles measuring 180°. $180 \times 4 = 720°$; the answer is the same as for part a.
    c) i) 120° ii) 60°
11. Answers may vary.
    a) 4:14; 6:21 b) 3:11; 9:33
    c) 7:3; 14:6 d) 25:6; 50:12
12. a) $\frac{3}{4}$ b) 54
13. 775 g for $5.49
14. 412.5 km
15. a) $39.00 b) $4.13 c) $86.20
16. a) $16.50 b) $566.50

Chapter 5 Graphing Relations

5.1 Interpreting Scatter Plots, page 148

1. a) 26.1°, 29.8° b) 40 m
   c) The scatter plot shows a downward trend; as the depth increases, the temperature decreases.
2. a) The average daily temperature in March for several Ontario cities and towns
   b) 2°C c) About 175 m
   d) There are no trends in the data.
3. a) i) 154 cm ii) 2 students
   b) i) 177 cm ii) 5 students
      iii) 2 students

c) There is an upward trend; as the height increases, the arm spans increase.
4. a) i) 14 h
   ii) 3
   iii) Yes, the student who earned a little less than $150 worked for 12 h; the coordinates of the point are approximately (12, 145).
   b) i) About $240
   ii) 20 h
   iii) Yes, the student who worked 21 h earned $200; the coordinates of the point are approximately (21, 200).
   c) There is an upward trend in the data; the points tend to go up to the right; as the time increases, the earnings increase.
5. a) i) $72   ii) $9/h
   b) The point (4, 48) represents a person who earns $12/h. The point (5, 40) represents a person who earns $8/h.

5.2 Line of Best Fit, page 153

1. Graph b; in this graph, the number of points above the line is close to the number of points beneath the line.
2. Answers will depend on the data students collected in *Investigate*.
3. a) *Temperature* against *Latitude*; there is no trend in the *Temperature* against *Elevation* scatter plot, so we may not draw a line of best fit.
   b) Answers may vary. For example: At a latitude of 44°N, the temperature is about 5.5°C.

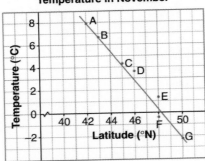

4. a) An upward trend; the rebound height increases as the drop height increases.
   b), c) Yes, the graph shows an upward trend.

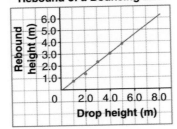

d) 1.8 m; 6.2 m.
e) Answers may vary. For example: From which height should the ball be dropped so its rebound height is 2.5 m? *Answer:* About 3.4 m
5. a) Yes, the taller the person, the greater her/his knee height.
   b), c), d) Answers will depend on students' data from *Investigate*.
6. Yes, the taller the person, the greater her/his shoe size.
   a), b) Answers will depend on students' data.
7. Answers will vary, depending on the measures chosen.
8. No, the number of points above and below the line will vary.
9. a) As the years increase, the record time decreases.

Year	Time (s)
1983	36.68
1983	36.57
1987	36.55
1988	36.45
1992	36.43
1993	36.41
1993	36.02
1994	35.92
1996	35.76
1998	35.39
1998	35.36
1999	34.76
2000	34.63
2001	34.32

b)

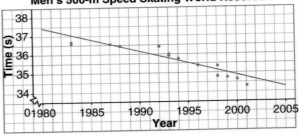

c) About 37.3 s. The actual world record in 1981 was 36.91 s.
d) Questions may vary. For example: What might the world record time be in 2005? *Answer:* About 34.2 s

5.3 Curve of Best Fit, page 160

1. a) As the time increases, the height increases, then decreases.

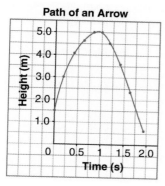

b) As the time increases, the number of animals increases.

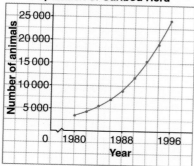

c) As the time increases, the height increases, then levels off.

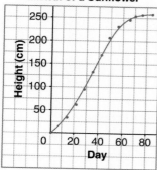

2. a) The number of daylight hours increases from January to July, then decreases from July to December.

b)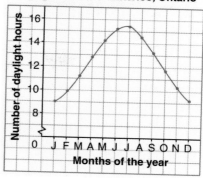

c) About 12.0 h   d) About 15.6 h

d) Answers will vary. From your birth date on the horizontal axis move up to the graph, then across to the vertical axis to find the number of hours of daylight.

3. a) As the time increases, the number of hosts increases. Each year, the number is about 3 or 4 times the number the year before, until about 2000, when the number doubles, then it is multiplied by 1.5 each year.

b)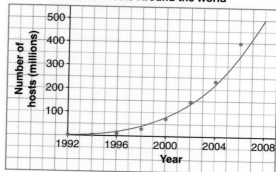

c) About 110 million
d) About 2008

4. a) As the time increases, the height decreases.

b)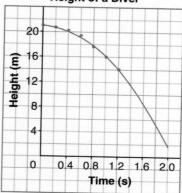

c) After about 1.5 s
d) Extend the graph until it reaches the *Time* axis: a little more than 2.1 s

ANSWERS 327

5. a)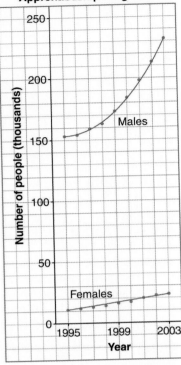

Enrollment in Apprenticeship Programs

b) Line of best fit females: each year, the number of people increases by about 1000 or 2000. Curve of best fit males: for most years, the increase is greater than the year before.

c) Extend the line and the curve to the year 2003: About 232 000 males, about 23 000 females

Technology: Using a Graphing Calculator to Draw a Curve of Best Fit, page 166

1. Quadratic curve

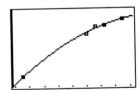

2. Exponential curve

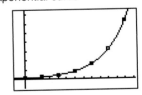

Chapter 5 Mid-Chapter Review, page 168

1. a) Each point represents a boy. The horizontal coordinate represents the time played in minutes and the vertical coordinate represents the points he scored.
   b) 7 points
   c) 7 boys, 2 boys
   d) In the top left corner of the graph
   e) As the time played increases, the number of points scored increases; the points go up to the right.

2. a) There is an upward trend indicating that the taller the dog, the greater its mass.
   b) Masses and Heights of Dogs

   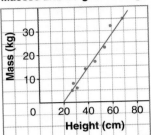

   c) Yes, the line of best fit goes up to the right; as the height increases, the mass increases.
   d) 42 cm   e) 22 kg
   f) Assume that the data for the dogs in parts d and e lie on the line of best fit.

3. a) The population increases as years go by; in later years, the population increases by a greater amount in each 10-year interval.
   b) World Population

   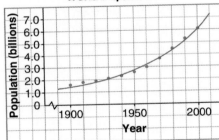

   c) About 2.0 billion, about 4.0 billion
   d) About 1.3 billion
   e) About 7.3 billion
   f) The data for the years in parts c, d, and e lie on the line of best fit.

5.4 Graphing Linear Relations, page 171

1. a) Yes, the relationship is linear because the points lie on a straight line.
   b) 77°F
   c) −20°C
   d) Use a thermometer that shows the temperatures in Fahrenheit and Celsius degrees
   e) Questions may vary. For example: What is the Celsius temperature for 50°F? *Answer:* 10°C

2. a) Yes, since the relationship for a pyramid is linear.

b) **Square Prisms with Base Area 9 cm²**

Height h (cm)	Volume V (cm³)
1	9
2	18
3	27
4	36
5	45

c)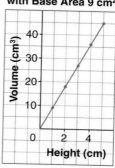

Volume of Square Prism with Base Area 9 cm²

d) As the height increases, the volume increases; the relationship is linear.
e) The volume is 9 times the height.
f) 31.5 cm³. Use the formula or the graph.

3. a) **Acute Angles in a Right Triangle**

One acute angle (°)	Other acute angle (°)
5	85
10	80
15	75
20	70
25	65

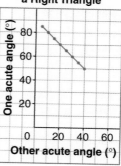

Acute Angles in a Right Triangle

b) As one angle increases, the other angle decreases by the same amount; the points lie on a straight line.
c) 60°, 50°

4. a)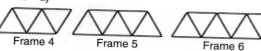
Frame 4 Frame 5 Frame 6

Frame number	Number of toothpicks
1	3
2	5
3	7
4	9
5	11
6	13

b) Yes, as the frame number increases by 1, the number of toothpicks increases by 2.
c) The relationship is linear; the points lie on a straight line.

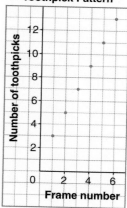

Toothpick Pattern

d) Add 2 more rows to the table; 17 toothpicks
e) The number of toothpicks is 1 more than two times the frame number.
f) Extend the pattern; or extend the table; or use the rule: 31 toothpicks

5. a) Yes, the length decreases by 1 cm, the width increases by 1 cm.
b) Yes, the points lie on a straight line.

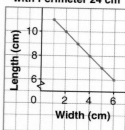

Dimensions of a Rectangle with Perimeter 24 cm

ANSWERS 329

c) The length plus the width of the rectangle is 12 cm.
d) i) 7.5 cm   ii) 1.5 cm
e) Use the rule or the graph.

6. a), b), c)

Celsius temperature (°C)	Fahrenheit temperature (°F)
−20	−10
−15	0
−10	10
−5	20
0	30
5	40
10	50
15	60
20	70
25	80
30	90

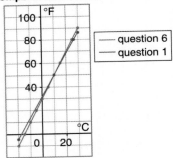

Relating Temperatures Scales

d) The rule "double and add 30" is useful from about 0°C to 20°C; beyond that, in either directions, the temperature the rule gives is more than 2°C different from the true temperature.

7. a) Square Pyramids with Base Area 16 cm²

Height (cm)	Volume (cm³)
1	5.$\overline{3}$
2	10.$\overline{6}$
3	16.0
4	21.$\overline{3}$
5	26.$\overline{6}$

b) Square Pyramids with Base Area 16 cm²

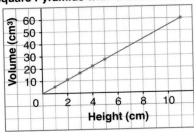

c) As the height increases, the volume increases. The relationship is linear because the points lie on a straight line.
d) Calculate or use the graph: 40 cm³
e) Extend the graph: 11.25 cm
f) The volume is one-third the height multiplied by 16.

5.5 Graphing Non-Linear Relations, page 177

1. a) Linear, because the points lie on a straight line
   b) Non-linear, because the points lie on a curve
2. a) i) 135 calories, about 202 calories
      ii) About 8 million, about 4 million
   b) Questions may vary. For example: By how much did the population increase from 1960 to 1980? *Answer:* About 3 million
3. a) Yes, because as the width increases by 1, the length decreases by a different amount each time.
   b) Yes, the points lie on a curve.

Dimensions of Rectangles with Area 36 cm²

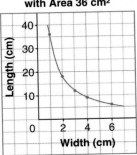

c) i) 7.2 cm   ii) 4.5 cm
d) Length times width is 36; or length is 36 divided by width; or width is 36 divided by length.

4. a)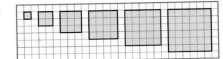

b)

Side length (cm)	Area (cm²)
1	1
2	4
3	9
4	16
5	25
6	36

c) As the side length increases by 1, the area increases by a greater amount each time; the increases are consecutive odd numbers: 3, 5, 7, 9, 11

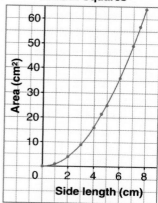

d) About 20 cm²
e) About 56 cm²
f) Multiply each side length by itself.

5. a) The edge length of each cube increases by 1 unit each time.

b)

Edge length (cm)	Number of cubes
1	1
2	8
3	27
4	64
5	125
6	216
7	343

c) As the edge length increases, the number of cubes increases; each increase is much greater than the preceding increase.

d) i) Multiply:
   edge length × edge length × edge length
   ii) The number of cubes is equal to the volume in cubic units.

f)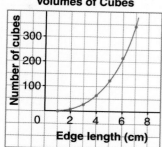

g) $V = e^3$ or $V = e \times e \times e$
h) 8
i) 1000
j) Extend the graph; or use the rule; or extend the pattern.

6. a) The graph will go down to the right. It will not be a straight line because the value of the car decreases by a different amount each time.

b)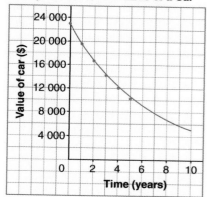

c) The graph goes down to the right; the points lie on a curve.
d) Extend the graph: between 7 and 8 years

7. Answers will vary depending on students' data; the mass does *not* affect the time to complete 6 swings.

8. Answers will vary depending on the measurements chosen.

9. Answers will vary. For example:
a) I estimate my total daily caffeine intake is 70 mg.

b)

Time (h)	Mass of caffeine (mg)
0	70
6	35
12	17.5
18	8.75
24	4.375

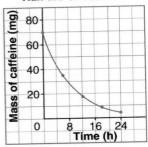

Half-life of Caffeine

c) Answers may vary. For adult family members, the initial data may be greater, but the graph has the same shape.

10. The relationship is non-linear.

5.6 Interpreting Graphs, page 182

1. a) Graph C b) Graph A c) Graph B
2. a) Walk at a constant speed toward the CBR for 3 s, stop for 4 s, then continue to walk toward the CBR at a slower speed.
   b) Walk away from the CBR at a steady speed for 3 s, stop for 4 s, then continue to walk at the same speed toward the CBR.
   c) Walk very slowly toward the CBR for 2.5 s, stop for 4 s, then walk away from the CBR at a faster speed.
3. From A to B: walks 200 m for 3 min
   From B to C: stops for 1 min
   From C to D: walks 300 m for 6 min
   From D to E: stops at the store for 7 min
   From E to F: walks 500 m home in 11 min
4. From 5:00 a.m. to 9:00 a.m., the volume decreases from 1 million litres to 500 000 L; people are having baths and showers, watering lawns, washing clothes and dishes.
   From 9:00 a.m. to 11:00 a.m., the volume remains constant, the water entering the reservoir is equal to that being used.
   From 11:00 a.m. to 2:00 p.m., more water enters the reservoir than leaves it, and the volume increases to 700 000 L.
   From 2:00 p.m. to 3:00 p.m., the volume is constant.
   From 3:00 p.m. to 10:00 p.m., the volume decreases to 300 000 L. People are home from work and school and use water at a faster rate than it enters the reservoir.
5. From 0 min to 7 min, the temperature remains the same. All the heat is used to melt the ice, so the temperature does not change.
   From 7 min to 17 min, the water is heated to its boiling point of 100ºC.
   From 17 min to 20 min, the water continues to boil, and the heat is used to create steam, so the temperature does not change.

Chapter 5 Review, page 186

1. a) How the price of a used car changes with age
   b) 3 years
   c) $9000 and $12 000; The $12 000 car may be in better condition, and have less mileage
   d) There is a downward trend; as the age of a car increases, its price decreases.
2. a), b) As the time increases, the height increases.

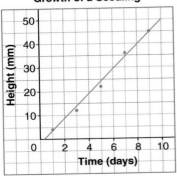

Growth of a Seedling

   c) i) About 18 mm ii) 50 mm
   d) About 12 days
   e) No, the seedling will probably not continue to grow at the same rate.
3. a) From 1996 to 1998, sales increased. From 1998 to 2001, sales decreased. From 2001 to 2003, sales decreased very slowly.
   b)

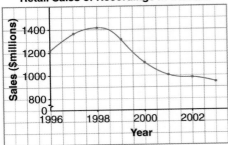

Retail Sales of Recordings in Canada

   c) About $1.4 billion, about $950 million
   d) About $940 million; I assume the point lies on the graph that moves down gradually to the right.

4. a) As time increases, the temperature decreases.

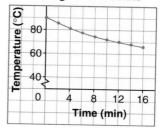

b) A curve of best fit
c) At 1 min, the temperature will suddenly fall, then continue to decrease. Graphs may vary.

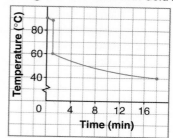

5. a) Frame 4

Frame 5

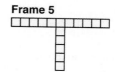

Frame 6

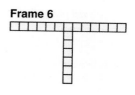

b)

Frame number	Number of squares
1	4
2	7
3	10
4	13
5	16
6	19

c) As the frame number increases by 1, the number of squares increases by 3. I think the graph will be linear.

d)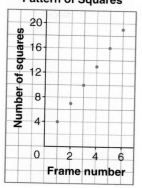

e) The number of squares is 1 more than 3 times the frame number.
f) Extend the table or graph; draw the next 2 frames; or use the rule: 25 squares
g) The 9th frame; I added 3 squares to the 8th frame to get the next frame.

6. a) As the resistance increases, the current decreases.

b)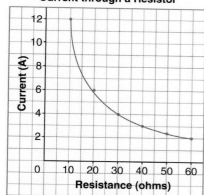

c) About 8 A
d) About 12 ohms

7. From A to B, Hasieba travels 200 m in 2 min. From B to C, she walks slower and travels 150 m in 4 min. From C to D, Hasieba stops for 4 min. From D to E, Hasieba walks 250 m in 3 min. From E to F, she stops for 3 min. From F to G, Hasieba walks 600 m home in 9 min.

Chapter 5 Practice Test, page 188

1. C
2. C

3. a) Line of best fit, because when the depth increases by 5 m, the increase in pressure is always 50 kPa.

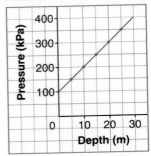

b) 12.5 m, 30 m   c) 100 kPa
d) As the depth increases, the pressure increases.
4. a) As the number of bounces increases, the height decreases.
b) Curve of best fit, because when the number of bounces increases by 1, the decrease in height changes.

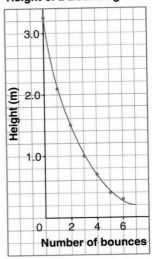

c) Extend the graph to the left to 0 bounces. The height is about 3.2 m.
d) Answers may vary. Extend the graph to the right to find the height of the 7th bounce, which is about 0.2 m.
5. If the points lie on a straight line, it is a linear relation. If the points lie on a curve, it is a non-linear relation. If the points are scattered on a grid and do not lie on a line or a curve, there is no relationship.
6. William has a head start of about 40 m. He runs for 10 s, slows down for 8 s, then runs for 12 s and slows down until he reaches the finish line at 40 s. Rhiannon runs for 12 s, slows down for 8 s, runs for 10 s, overtakes William, and wins the race in 34 s.

Chapter 6 Linear Relations

6.1 Recognizing Linear Relations, page 194

1. a) 6, 6, 6, 6; linear relation because all the first differences are equal
b) 1.0, 0.8, 0.7, 0.7; non-linear relation because the first differences are not equal
c) 5, 5, 5, 5; linear relation because all the first differences are equal
2. Answers may vary.
a) The cost to rent a canoe at $6/h
b) The mass of a baby from 0 to 4 months
c) The distance driven at an average speed of 15 m/s
3. a)

Frame number	Area (square units)	First differences
1	2	
		4
2	6	
		6
3	12	
		8
4	20	
		10
5	30	
		12
6	42	

b) See the table in part a. The relation is not linear because the first differences are not equal.
c) Yes, the points lie on a curve.

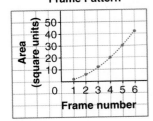

4. a)

Frame number	Number of tiles	First differences
1	1	
		2
2	3	
		2
3	5	
		2
4	7	
		2
5	9	
		2
6	11	

b) See the table in part a. The relation is linear because the first differences are equal.
c) Yes, the points lie on a straight line.

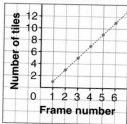

d) Extend the graph or the table: 17 tiles

5. a) −4, −4, −4, −4; the volume of water in litres that leaks out every hour
b) The relation is linear because the first differences are equal.
c) Yes, the points lie on a straight line.

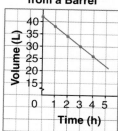

d) 22 L; I assume water continues to leak at the rate of 4 L/h.
e) Answers may vary. For example: When is there 20 L of water in the barrel? *Answer:* After 5.5 h

6. a)

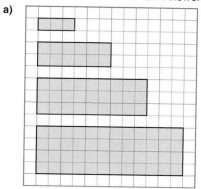

b) 9, 15, 21
c) The relation is non-linear because the first differences are different.

d) The graph is a curve that goes up to the right.

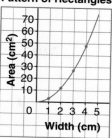

e) 75 cm^2

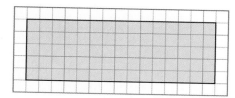

7. a) 5
b) 125, 125, 125, 125; the distance in metres for 5 lengths of the pool
c) The relation is linear because the first differences are equal.
d) Yes, the points lie on a straight line.

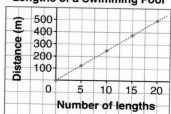

e) 25 m

8. Answers will depend on the pattern drawn.

6.2 Average Speed as Rate of Change, page 199

1. a) 80 km/h   b) 15 km/h
2. a) The average speed of the car
   b) The average speed of the cyclist
3. The rate of change is 75 km/h.
4. a)

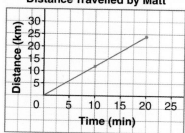

b) 1.2 km/min

c) 1.2 km/min; the average speed is equal to the rate of change
d) 72 km/h

5. a)

Time (min)	Distance (km)
0	0
30	2
60	4
90	6
120	8
150	10
180	12
210	14
240	16

Distance Hiked by a Group

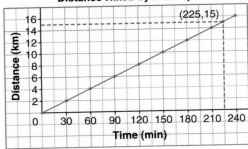

b) $0.0\overline{6}$ km/min or $\frac{1}{15}$, the distance the hikers walk in 1 min
c) Use the graph or table, 225 min
d) 4 km/h
e) Answers may vary. For example: How far have hikers travelled after 2.5 h? *Answer:* 10 km

6. Make a table of values; draw a graph; calculate each average speed: Keta because her average speed is 10.5 km/h.

6.3 Other Rates of Change, page 203

1. a) 15 km/L   b) $25 per guest
2. a) The distance driven on 1 L of gas
   b) The cost for 1 guest
3. Smart car: 792 km; SUV: 720 km
4. a) **How a Spring Stretches**

b) Yes, the points lie on a straight line through the origin.
c) 0.4 cm/g; for every increase in mass of 1 g, the spring stretches 0.4 cm.

5. Answers will vary depending on the quantities chosen; for example, the rate of change could represent the average speed in kilometres per hour.
   a), b) Cost of party on the vertical axis; number of guests on the horizontal axis
   c) The rate of change is $5 per guest; this means that each guest costs $5 after $100 has been paid to rent the hall.

6. It will take 12 min in total, or 8 min from when the plane is at 1200 m

7. a) Approximately 2300 m and 1300 m
   b) Approximately –143 m/min
   c) The height is decreasing at a rate of 143 m per minute.

8. a) For the horizontal portion of the graph, the rate of change is 0. For the sloping portion of the graph, the rate of change is $1.25/person.
   b) The cost is constant at $20 for any number of people from 0 to 10.
   For 11 or more people, the cost increases by $1.25 per person.
   c) $2.50

6.4 Direct Variation, page 207

1. a) No, the graph does not pass through origin.
   b) Yes, the graph passes through the origin.
   c) Yes, the graph passes through the origin.
   d) No, the graph is not a straight line.

2. a)

Side length (cm)	Perimeter (cm)
0	0
1	4
2	8
3	12
4	16
5	20

b) 4 cm/cm

Perimeter of a Square

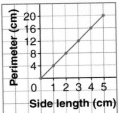

c) The perimeter is 4 times the side length.
d) 44 cm

3. a) This is a direct variation situation because the graph is a straight line through the origin

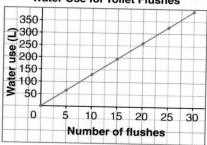

b) 13 L/flush; one toilet flush uses 13 L of water.
c) The water use in litres is 13 times the number of flushes.
d) 221 L

4. a)

Mass of bird seed (kg)	Cost ($)
1	0.86
2	1.72
3	2.58
4	3.44
5	4.30
6	5.16

b) Yes, the graph is a straight line passing through the origin.

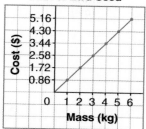

c) The cost in dollars is 0.86 times the mass in kilograms.
d) $11.18
e) Yes, when 2 quantities are related by direct variation, when you double one quantity, the other quantity also doubles: the cost is $22.36.

5. a)

Number of trees, n	Earnings, E ($)
0	0
100	16
200	32
300	48
400	64
500	80
600	96
700	112
800	128
900	144
1000	160

b) Yes. The graph is a straight line through the origin.

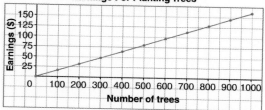

c) $0.16 per tree; the earnings for planting 1 tree
d) $E = 0.16n$ where E is the earnings in dollars and n is the number of trees planted
e) $272

6. a) Yes, the graph is a straight line through the origin.

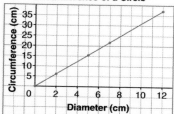

b) 3.1; the increase in circumference when the diameter increases by 1 cm; it is an approximate value of π.
c) $C = 3.1d$, where C is the circumference and d is the diameter.
d) About 40.3 cm

ANSWERS 337

7. a) Yes, the graph is a straight line through the origin.

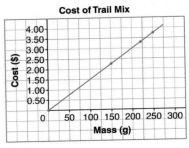

b) $0.015/g; the cost of 1 g of trail mix
c) $C = 0.015m$
d) $5.10
e) Questions may vary: How much trail mix could you buy with $4.00? *Answer:* About 270 g

8. Answers may vary. For example:
   a) The cost in dollars of v kilograms of apples that cost $0.90 for 1 kg
   b) The distance in kilometres driven at 90 km/h for t hours
   c) The cost in dollars for n books that each cost $5.50

9. Answers may vary.
10. a) We cannot tell, because Anastasia might have stopped for a coffee on the way to work.
    b) 44 km/h

6.5 Partial Variation, page 213

1. a) Yes, the graph is a straight line not passing through the origin.
   b) No, the graph is not a straight line.
   c) Yes, the graph is a straight line not passing through the origin.
   d) No, the graph passes through the origin so it represents direct variation.
2. a) $45; $1/subscription
   c) 3000 m; –400 m/lap
3. a)

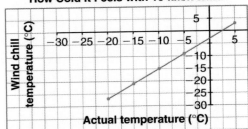

   b) Yes, the graph is a straight line not passing through the origin.
   c) About –19°C, about 1°C

4. a) Join the points with a solid line because all altitudes and temperatures are possible.

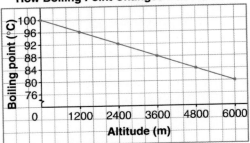

   b) Yes, the graph is a straight line not passing through the origin.
   c) i) About 90°C
      ii) About 83°C

5. a)

Number of guests	Cost ($)
0	500
10	750
20	1000
30	1250
40	1500
50	1750

   b) Join the points with a broken line because a fraction or decimal number of guests is not possible.

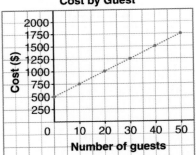

   c) Yes, the graph is a straight line not passing through the origin.
   d) $25/person; the extra cost for each extra person
   e) $1625, $2750
   f) No, the situation is not represented by direct variation, so the cost is not proportional to the number of guests.

6. a) $C = 20 + 0.10n$
   b) $38.00
7. a) $h = 25 + 1.3n$
   b) 32.8 cm, 37.35 cm

c) Questions may vary. How much did Nuri's hair grow in 3 months? *Answer:* 3.9 cm
8. Answers may vary.
   a) i) The cost in dollars of a party when the hall rental costs $150 and the cost per person is $20
      ii) The length in centimetres of a rectangle when the length is 20 times the width
      iii) The cost in dollars for *m* items, when 1 item costs $0.75
   b) Equations ii and iii represent direct variation because their graphs are straight lines that pass through the origin.
   Equation i represents partial variation because its graph is a straight line that does not pass through the origin.
9. a) **Cost Against Number of Posters**

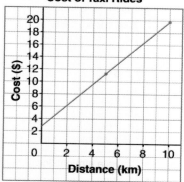

   b) $50; it is the vertical intercept.
   c) $0.25/poster; the cost per poster after the fixed cost has been paid
   d) $C = 50 + 0.25p$, where C is the cost in dollars for p posters sold
   e) $112.50
10. a) **Cost of Taxi Rides**

   b) $2.75; it is the vertical intercept.
   c) $1.70/km; the cost per kilometre after the fixed cost has been paid
   d) $C = 2.75 + 1.70d$, where C is the cost in dollars for a ride that is d kilometres
   e) $28.25
   f) The flat rate fare (the partial variation fare is $165.95)

11. a) i) Yes, the number of points is a fixed number plus a number that depends on the number of items handed in
      ii) $P = 5i + 20$, where P is the total number of points when i items are handed in within 45 min
   b) i) 41 or more items
      ii) 45 or more items

Chapter 6 Mid-Chapter Review, page 220

1. a) The height increase in 1 min
4. a)

Time (min)	Height (m)	First differences
0	0	
1	200	200
2	400	200
3	600	200
4	800	200

   b) Yes, all first differences are equal.
   c) The rate of change is 200 m/min; it has the same numerical value as the first differences.

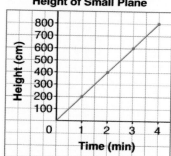

   d) 8 min from when the plane took off
2. a)

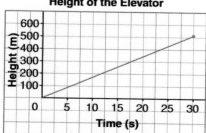

   b) i) 255 m
      ii) 340 m
   c) 17 m/s
   d) 17 m/s

3. a)

Rental time (days)	Cost ($)
0	0
2	66
4	132
6	198
8	264
10	330

b)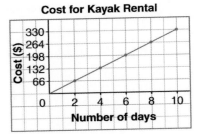

c) Direct variation because the graph is a straight line passing through the origin
d) $33/day; the daily cost to rent a kayak

4. a)

Mass of turkey (kg)	Cooking time (min)
0	15
2	75
4	135
6	195
8	255
10	315

b) 165 min

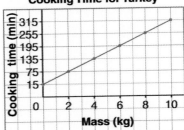

c) Cooking time is 30 min plus 30 min times the mass in kilograms
d) $t = 30m + 15$
e) 231 min

6.6 Changing Direct and Partial Variation Situations, page 222

1. a) Direct variation; there is no fixed or initial cost.
   b) Partial variation; there is a fixed cost plus a cost that depends on the number of brochures printed.
   c) Partial variation; there is a fixed amount plus an amount that depends on the number of subscriptions sold.

2. a) Partial variation
   b) Direct variation
   c) Direct variation
   d) Partial variation

3. a) Direct variation
   b)

Time (min)	Cost ($)
0	0
10	2.00
20	4.00
30	6.00
40	8.00
50	10.00
60	12.00
70	14.00
80	16.00
90	18.00
100	20.00

c), e)

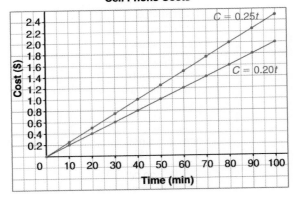

d) $C = 0.20t$
e) i) The new graph is steeper than the original graph; the rate of change increases from $0.20/min to $0.25/min.
   ii) The rate of change increases as in part i: $C = 0.25t$

4. The second pay scale is better because Jocelyn earns more money when she sells 2 or more ads.

5. a)

Time (min)	Cost ($)
0	12.00
10	13.00
20	14.00
30	15.00
40	16.00
50	17.00
60	18.00
70	19.00
80	20.00
90	21.00
100	22.00

b), d)

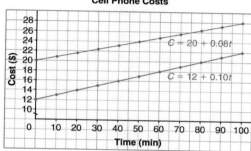

c) $C = 12 + 0.10t$
d) i) The new graph is less steep and its vertical intercept is greater.
   ii) $C = 20 + 0.08t$

6. For the plan in question 3, the cost doubles when the time doubles because the plan is represented by direct variation and the cost is proportional to the time. For the plan in question 5, the cost does not double when the time doubles because the plan is represented by partial variation and the cost is *not* proportional to the time.

7. a)

Distance (km)	Cost ($)
0	29.00
50	29.00
100	29.00
150	34.00
200	39.00
250	44.00
300	49.00

b)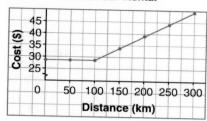

c) Neither; the graph is not a straight line.
d) The graph has 2 segments: one horizontal segment that represents a fixed cost of $29 up to 100 km, and a sloping segment that shows how the cost changes as the distance increases.
e) $29.00, $36.50, $47.00

6.7 Solving Equations, page 229

1. a) $x = 4$ b) $x = 9$ c) $x = 18$
   d) $x = 2$ e) $x = 1$ f) $x = 9$
2. a) $x + 4 = 9$; $x = 5$
   b) $2x + 5 = 13$; $x = 4$
   c) $3x = 18$; $x = 6$
   d) $6 = x + 3$; $x = 3$
   e) $11 = 2x + 3$; $x = 4$
   f) $2x = 8$; $x = 4$
3. a) $x = 5$ b) $x = 3$ c) $x = 7$
   d) $x = 3$ e) $x = 2$ f) $x = 8$
4. a) $x = -1$ b) $x = -1$ c) $x = -1$
   d) $x = -3$ e) $x = -2$ f) $x = -3$
   g) $x = -10$ h) $x = -4$
6. a) 24 cm   b) 10 cm
   c) 15 cm
7. a) 86°F   b) −10°C
8. 206 barrels
9. a) i) $2x + 40 + 70 = 180$
      ii) $5x + 20 = 50$
      iii) $58 + 58 + 3x + 7 = 180$
   b) i) $x = 35$
      ii) $x = 6$
      iii) $x = 19$

6.8 Determining Values in a Linear Relation, page 233

1. a) i) $48   ii) $96
   b) i) $80   ii) $240   iii) $400
2. a)

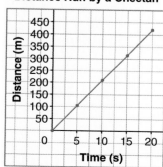

Distance Run by a Cheetah

b) About 19 s
c) About 24 s; assume cheetah maintains an average speed of 21 m/s
d) $d = 21t$
e) About 19 s; about 23.8 s
Using the formula is easier and more precise than using the graph.

3. a) 15 min

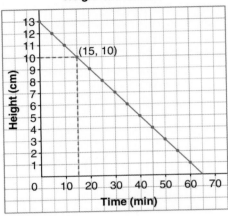

b)

Time (min)	Height (cm)
0	13.0
5	12.0
10	11.0
15	10.0
20	9.0

c) Answers may vary. For example: The table is the easiest way.

4. a) After 25 games

Number of games played, n	Value of card, V ($)
1	35.00
2	28.00
3	21.00
4	14.00
5	7.00
6	0

b) After 25 games

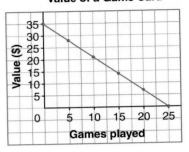

c) $V = 35 - 1.40n$
d) 25 games
e) Answers may vary.

5.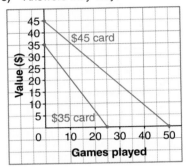

a) The vertical intercept changes from $35 to $45. The graph is less steep.
b) 50 games in total, so 25 more games
c) Answers may vary. For example: The $45 card is the better deal.

6.9 Solving Problems Involving Linear Relations, page 237

1. a)

Distance travelled (km)	Fare ($)
0	2.00
1	3.50
2	5.00
3	6.50
4	8.00
5	9.50

b)

Time (months)	Amount ($)
0	900
1	750
2	600
3	450
4	300
5	150
6	0

c)

Dollars	Pesos
0	0
20	180
40	360
60	540
80	720
100	900

2. a) Partial variation because the points lie on a line not passing through the origin.
   b)

Frame	Toothpicks
1	6
2	11
3	16
4	21
5	26
6	31
7	36

   c) $T = 5n + 1$ d) 86 toothpicks

3. a)

Time (h)	Car A distance (km)	Car B distance (km)
0	0	0
0.5	45	30
1	90	60
1.5	135	90
2	180	120

   b) Car A: $d = 90t$; car B: $d = 60t$
   c) Car A; the average speed is the rate of change.
   d) Car A: 157.5 km; car B: 105 km

   e) Car A: About 1.2 h; car B: about 1.8 h
4. $800
   a)

Sales, s ($)	Earnings, E ($)
1000	550
2000	600
3000	650
4000	700
5000	750
6000	800

   b) Use $E = 500 + 0.05s$; $800
   c) Answers may vary.   d) Drawing a graph
5. a) 20 min
   b) Use: a graph; an equation; a table of values
6. 25 min
7. a) $d = 3 + 2t$
   b) Answers may vary. For example: The distance a ball travels at an average speed of 2 m/s, and the ball has rolled 3 m before the time is measured.

6.10 Two Linear Relations, page 242

1. a) About 290 m
   b) About (5.5, 250); the two balloons meet after 5.5 min at a height of about 250 m.
2. a) 4000 m or 4 km
   b)

Time (min)	Nyla's distance (m)	Richard's distance (m)
0	4000	0
5	3500	1500
10	3000	3000
15	2500	4500
20	2000	6000
25	1500	7500
30	1000	9000

Nyla and Richard's Distances

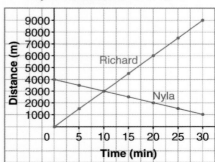

ANSWERS 343

c) (10, 3000); Nyla and Richard meet after 10 min, when Richard has travelled 3000 m, and Nyla is 3000 m from where Richard started.

3. a)

Laps	Sophie's pledges ($)	Hatef's pledges ($)
0	96	0
10	146	130
20	196	260
30	246	390
40	296	520
50	346	650

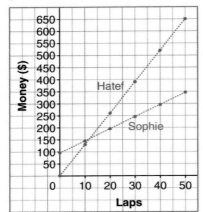

b) Hatef: $650; Sophie: $346
c) (12, 156); when both people have swum 12 laps, each has earned $156 in pledges.

4. a) $C = 150 + 0.25n$ b) $R = 2.25n$
c)

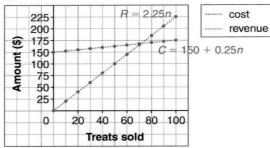

d) 75 treats

5. a) In 2.5 h; each person has read 100 pages.
b) Answers may vary; table of values; graph

Chapter 6 Review, page 248

1. a)

Frame 4

b)

Frame number	Number of squares	First differences
1	1	
		3
2	4	
		5
3	9	
		7
4	16	
		9
5	25	
		11
6	36	

c) Non-linear relationship because the first differences are not equal
e) The points lie on a curve so the relationship is non-linear.

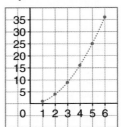

2. a) 70 km/h; average speed of a car
b) −2 L/h; the rate at which water evaporates

3. a) Direct variation because the graph is a straight line passing through the origin
b) Partial variation because the graph is a straight line *not* passing through the origin

4. a)

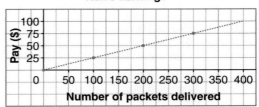

b) $62.50
c) 400 packets
d) 560 packets

5. a) **Cost for Renting a Snowboard**

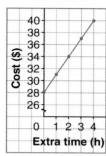

b) The rental cost when a snowboard is rented for 8 h
c) $3/h; the extra cost ($3 per hour) for renting a snowboard beyond 8 h
d) $C = 3t + 28$

6. a)

Time (min)	Volume of air (m³)
0	0
1	450
2	900
3	1350
4	1800
5	2250
6	2700

b), d) **Volume of Air in a Balloon**

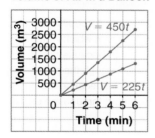

c) $V = 450t$
d) i) The new graph is less steep.
   ii) $V = 225t$

7. a) $x = 1$ b) $x = -4$ c) $x = 4$
   d) $x = -3$ e) $x = 3$ f) $x = -5$

8. a) Cost in dollars is 6 plus 0.75 times the number of toppings.
   b) $C = 6.00 + 0.75n$
   c) 5 toppings; use a table, a graph, or the equation

9. a) 27 cm   b) 140 cm
   c) $L = 0.15h$, where L is the length of the foot in centimetres and h is the height in centimetres
   d) Make a table; draw a graph.

10. a) Speed Dot Company: $C = 2.40t$
    Communications Plus: $C = 12 + 0.90t$
    b)

Time (h)	Cost ($) Speed Dot Company	Cost ($) Communication Plus
0	0.00	12.00
2	4.80	13.80
4	9.60	15.60
6	14.40	17.40
8	19.20	19.20
10	24.60	21.00

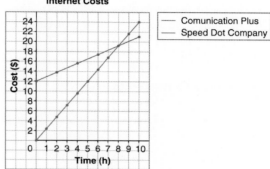

c) (8, 19.20); for 8 h of use, both plans cost the same amount, $19.20
d) Jo-Anne should choose Speed Dot Company because it is cheaper for 7 h.

Chapter 6 Practice Test, page 250

1. C
2. D
3. a) **Andy's Earnings**

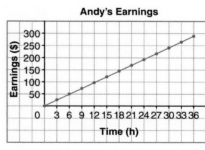

b) $8/h   c) $288   d) 27 h

4. Answers may vary. For example:
The cost of hall rental before any guests are included ($250)
The cost for each guest, above the hall rental ($15/person)
The cost for 10 guests ($400); 20 guests ($550); 30 guests ($700); 40 guests ($850); 50 guests ($1000)

5. $80
   a) **Cost for Placing an Ad**

   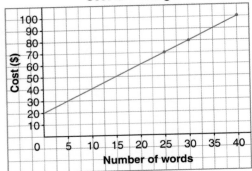

   b) $C = 20 + 2a$; the ad will cost $80.
   c) Answers may vary.
6. For 190 brochures, the costs are equal ($470).
   For less than 190 brochures, printer B is cheaper.
   For more than 190 brochures, printer A is cheaper.

Chapter 7 Polynomials

7.1 Like Terms and Unlike Terms, page 255

1. a) $2x^2$ b) $2x + 3$
   c) $-2x + 4$ d) $-x^2 + 3x - 2$
   e) $4x^2 + 6x - 1$ f) $-3x^2 - x - 2$
2.
3. $1, x, x^2$; the name of the tile describes its area.
4. $2x$ and $-4x$ are like terms.
   They are modelled by like tiles.

5. $3x^2$ and $3x$ are unlike terms. They are modelled by different sized tiles.

6. a) $-5x, 4x, -3x, 7x$ b) $-3x^2, x^2, 3x^2$
   c) Like terms have the same variable raised to the same exponent.
7. a) $x + 4$ b) $2 + x$ c) $-2x^2 - x - 1$
   d) $3x^2 + x + 1$ e) $2x + 4$ f) $x^2 + 2x + 1$
8. a) $5x + 4$ b) $7x^2 + 3x$ c) $x^2 + x + 4$
9. Answers may vary. For example: $x - 4x + 3 + 5x - 1$, which simplifies to $2x + 2$.

10. a) i) $2x + 1$ ii) $6x^2 - 9x$
    iii) $4x^2 + 6x + 3$ iv) $-3x^2 + x + 1$
    b) Answers may vary. For example: $-5x^2 - 1$
    There are no like terms.
11. a) x^3
    b) x^3 cube
    c) i) $8x^3$ ii) $4x^3 + 8x$ iii) $3x^3 + 3x^2 + 5$
12. a) To ensure it has a different area. The y-tile represents a variable different from x.
    b)
    c) $-5x - y + 3$. There are no like terms.

7.2 Modelling the Sum of Two Polynomials, page 259

1. a) $-x + 5$ b) $3x^2 + 2x - 3$
2. a) $-2x^2 - 3$

   b) $2x^2 + 3$

   c) $-2x^2 + 3$

   d) $-4x^2 + 7$

3. a) $6x + 2$ b) $3x + 2$ c) $4x + 1$
   d) $-x - 2$ e) $2x - 8$ f) $2x + 4$
4. a) $8x^2 + 3x - 1$ b) $2x^2 - 12x - 5$
   c) $4x^2 - x + 1$ d) $-x^2 + 2x - 8$
   e) $-5x^2 + 2x + 4$ f) $x^2 + 5x - 3$
5. $x^3 - 4x^2 + 3x - 2$. I combined like terms.
6. a) $5x + 9$ b) $18x + 16$
   c) $8x + 12$ d) $8x + 4$
7. a) Choose part a: 49   b) 49
   c) Evaluate after simplification because there is less arithmetic to do.
8. a) $2x^2 + 2x + 2$ b) $x^2 + x + 1$
   c) 0   d) $-x^2 - x - 1$
   e) $-2x^2 - 2x - 2$ f) $-2x^2 - 3x - 3$
9. For example, $(x^2 + x + 1) + (2x^2 - 3x + 4)$. There is no limit to the number of polynomials that can be found. Algebra tiles help.
10. $-x^2 - 2x + 5$. I used algebra tiles.

7.3 Modelling the Difference of Two Polynomials, page 265

1. $x^2 - x + 1$
2. a) $2x + 1$ b) $2x + 5$
   c) $8x + 1$ d) $8x + 5$
3. a) $x^2 + x + 3$ b) $x^2 - x - 5$
   c) $5x^2 - 3x - 3$ d) $x^2 - x + 3$
4. a) $-3x + 2$ b) $2x^2 - 3x$

   c) $-4x^2 + 7x - 6$ d) $x^2 - 5x + 4$

5. a) 2   b) 12
   c) $2x + 2$ d) $2x + 12$
6. a) $x^2 - 4$ b) $x^2 + x$
   c) $-1 - 2x^2$ d) $3x - 9x^2$
7. a) $2x^2 + 3x - 6$ b) $-2x + 2$
   c) $x^2 + 4x - 7$ d) $-2 + 2x - 2x^3$
   e) $3 + 6x - 5x^3$ f) 0
8. a) John did not properly subtract each term of the second polynomial.
   b) $-x^2 - 6x + 10$
   c) Check by adding $(3x^2 + 2x - 4)$ to the answer to obtain $2x^2 - 4x + 6$.
9. a) i) $x^2 + 4x$ ii) $-x^2 - 4x$ iii) $-12 + 2x^3$
       iv) $12 - 2x^3$ v) 0   vi) 0
   b) The answers in each pair are opposite polynomials.
   c) Answers may vary. For example:
      $(x + 1) - (3x^2 + 3) = -3x^2 + x - 2$
      $(3x^2 + 3) - (x + 1) = 3x^2 - x + 2$
      The answers are also opposite polynomials.
10. a) i) For example, $3x + 4$ ii) $-3x - 4$
    b) $6x + 8$. It is the polynomial in part a) i) added to itself.
    c) There is a pattern. Subtracting the opposite polynomial is the same as adding the opposite of the opposite polynomial, which is the original polynomial. The result will always be the original polynomial added to itself.
11. Answers may vary. For example:
    $(x^2 + 6x - 6) - (3x^2 + 2x - 1)$
    There is no limit to the number of possible polynomials.

Chapter 7 Mid-Chapter Review, page 268

1. No. The order in which the terms are written does not matter.
2. a) $4x - 2$ b) $1 - x^2$
   c) $2x^2 + 3x - 2$ d) $-4x^2 + 5x - 3$
3. a) $8x + 6$ b) $2x + 2$
   c) $3x^2 - 3$ d) $-2 - x$

   e) $-4x + 2$ f) $5x^3 + 2x$
   g) $-x^2 + x - 2$
4. No. The terms are unlike.

   $5x + 2$ $7x$

5. Answers may vary.
   For example: $3x + 6x - 5x + 4x$, which simplifies to $8x$. I used only like terms.
6. $x^2 + 2x - 1$
7. a) $8x + 9$ b) $2x + 2$
   c) $10x^2 - 4x - 2$ d) $6x^2 - x - 4$
   e) $-x^2 + 3x - 1$ f) $-x^2 + x - 3$
8. $2x^2 + 2x + 3$
9. a) $5x + 7$ b) $12x + 8$
10. a) $-x - 3$ b) $-3x^2 + 5x$
    c) $-2x^2 - 3x - 7$ d) $5x^2 + 2x + 1$
11. a) 7   b) $6x^2 - 2x$
    c) $-2x^2 + x + 9$ d) $-5x^2 - 6x + 2$
    e) $3x + 7$

7.4 Modelling the Product of a Polynomial and a Constant Term, page 271

1. a) $2(2x + 2) = 4x + 4$
   b) $3(2x + 1) = 6x + 3$
   c) $2(3x + 2) = 6x + 4$
2. a) $8x + 2$

   b) $6x + 2$

   c) $10x + 15$

   d) $16x + 12$

3. a) $3(2x + 7) = 6x + 21$
   b) $3(1 + 4x) = 3 + 12x$
   c) $5(2x + 9) = 10x + 45$
   d) $5(x + 8) = 5x + 40$
4. a) The coefficient of x increases by 2 and the constant term increases by 1 each time.
      $12x + 6$; $14x + 7$
   b) The constant term increases by 1 and coefficient of x increases by 2 each time.
      $6 - 12x$; $7 - 14x$

ANSWERS   347

5. $-400x^2 - 100x + 400$; Algebra tiles are not useful because you would need too many tiles.
6. Use CAS. The screen shows $3x^2 + 12x - 6$. Jessica forgot to multiply every term by 3.
7. a) $21x - 7$ b) $12x - 15$ c) $12x - 8$
   d) $25x^2 - 15x$ e) $-8x^2 + 12$
   f) $9x^2 + 9x - 54$
8. a) $-8x - 4$ b) $-15x^2 + 9$ c) $-2x - 8$
   d) $-7x^2 + 35$ e) $10x^2 - 2x$ f) $18 - 12x$
9. a) $-2x^2 + 4x - 8$ b) $3x^2 - 9x + 21$
   c) $8x^3 - 4x^2 - 2x$ d) $24x^3 + 16x^2 - 24$
   e) $-12x^2 + 6x - 30$ f) $4x^2 - 12x - 12$
   I used paper and pencil and the distributive law.
10. a) Square A: $P = 4(4x + 1) = 16x + 4$
    Square B: $P = 4 \times 3 \times (4x + 1) = 48x + 12$
    b) $32x + 8$
11. I would start by modelling the multiplication by the opposite (positive) constant.
    Then, I would replace each tile by its opposite and write the result.
    For example: $-2(x + 1)$. First model $2(x + 1)$

Replace each tile by its opposite to model $-2(x + 1)$.

$-2(x + 1) = -2x - 2$

7.5 Modelling the Product of Two Monomials, page 277

1. a) 12 b) $12x$ c) $12x^2$
2. All have coefficient 12. The first product has no factor of x, each of the other products has a different power of x.
3. a) $2x^2$ b) $4x^2$ c) $6x^2$
4. a) $3x^2$ b) $5x^2$ c) $15x^2$ d) $4x^2$
   e) $4x^2$ f) $12x^2$ g) $14x^2$ h) $9x^2$
5. a) The coefficients increase by 2 each time: $12x^2$, $14x^2$, $16x^2$
   b) The coefficients increase by 3 each time: $18x^3$, $21x^3$, $24x^3$
6. a) $6x^3$ b) $4x^3$ c) $3x^3$ d) x^3
   e) $24x^3$ f) $16x^3$ g) $24x^3$ h) $9x^3$
7. a) $-6x$ b) $-16x$ c) $-8x$ d) $-35x$
   e) $-18x$ f) $-10x$ g) $-81x$ h) $18x$
8. a) $-6x^3$ b) $-12x^3$ c) $20x^3$ d) $-x^3$
   e) $-32x^3$ f) $-4x^3$ g) $-12x^3$ h) $36x^3$
9. Use algebra tiles.

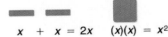

$x + x = 2x$ $(x)(x) = x^2$

10. a) i) $15x^2$ ii) $-15x^2$ iii) $15x$
    iv) $15x$ v) $15x^2$ vi) $-15x^2$
    vii) $15x^2$ viii) $15x^3$
    b) All the products in part a except viii) $15x^3$. An x^3 term cannot be modelled with algebra tiles.

11. Eighteen different pairs of monomials are possible:
    $(1)(36x^3)$; $(x)(36x^2)$; $(x^2)(36x)$; $(x^3)(36)$; $(2)(18x^3)$;
    $(2x)(18x^2)$; $(2x^2)(18x)$; $(2x^3)(18)$; $(3)(12x^3)$;
    $(3x)(12x^2)$; $(3x^2)(12x)$; $(3x^3)(12)$; $(4)(9x^3)$; $(4x)(9x^2)$;
    $(4x^2)(9x)$; $(4x^3)(9)$; $(6)(6x^3)$; $(6x)(6x^2)$

7.6 Modelling the Product of a Polynomial and a Monomial, page 281

1. a) $x^2 + 3x$ b) $2x^2 + 6x$
   c) $2x^2 + x$ d) $4x^2 + 2x$
2. a) $2x(3x + 4) = 6x^2 + 8x$
   b) $x(2x + 3) = 2x^2 + 3x$
   c) $3x(2x + 1) = 6x^2 + 3x$
   d) $3x(x + 4) = 3x^2 + 12x$
3. a) $3x^2 + 2x$ b) $4x^2 + 6x$
   c) $6x^2 + 3x$ d) $12x^2 + 16x$
4. a) $4x^2 + 12x$ b) $2x^2 + 3x$
   c) $2x^2 + 12x$ d) $15x^2 + 6x$
5. a) Each first term is x^2. The coefficients of x increase by 1 each time.
   b) $x^2 + 7x$; $x^2 + 8x$
6. Use an area model.
   The area of the rectangle is $x^2 + x$.

7. a) $12x^2 - 4x$ b) $8x^2 - 2x$ c) $5x^2 - 35x$
   d) $-4x^2 + 3x$ e) $6x^3 - 8x$ f) $-8x^2 - 12x$
   g) $6x^2 - 12x$ h) $-x^2 - x$
8. a) $-2x^2 - 6x$ b) $-2x^2 - x$ c) $2x^2 + 7x$
   d) $-6x + 3x^2$ e) $-28x^2 + 63x$ f) $-2x^2 - 3x$
   g) $-6x^2 + 10x$ h) $-12x^2 - 18x$
9. a) $2x^3 + 4x^2$ b) $-9x^3 + 15x^2$ c) $32x^3 + 24x$
   d) $14x^3 + 21x^2 - 7x$ e) $-8x^3 + 10x^2 - 14x$
   f) $12x^3 - 9x^2 - 27x$ g) $-2x^2 + 6x^3 - 2x$
   h) $x^3 - 9x + x^2$
10. $5x(9x + 7) - x(6x + 4) = 39x^2 + 31x$

7.7 Solving Equations with More than One Variable Term, page 285

1. a) $x = 1$ b) $x = 3$ c) $x = 4$
   d) $x = 2$
2. a) $x = 7$ b) $x = -4$ c) $x = -6$
   d) $x = -2$ e) $x = 1$ f) $x = 8$
3. a) $x = -7$ b) $x = 4$ c) $x = -1$
   d) $x = 6$ e) $x = -1$ f) $x = 2$
   g) $x = -1$ h) $x = 49$

4. Substitute to check. a) L.S = −17; R.S. = −17
   b) L.S. = 31; R.S. = 31
5. a) $x = 8$ b) $x = 2$ c) $x = 1$
   d) $x = -3$ e) $x = 21$ f) $x = 7$
6. a) $x = 2$ b) $x = -13$ c) $x = -2$
   d) $x = -3$ e) $x = 1$ f) $x = 15$
7. $x = 5000$. The number of parts that must be made and sold for the business to break even
8. a) $x = 12$ b) $x = -1$ c) $x = 2$

Chapter 7 Review, page 288

1. x has different exponents in each monomial.
2. a) $-5x + 3$ b) $2x^2 + 4x - 2$
   c) $3x^2 + 3x - 1$
3. a)

   b)
4. a) $7x + 7$ b) $7x + 6$ c) $3x^2 - 5x$
   d) $3 + x$ e) $-3x^3 + 3$ f) $x^3 + 4x$
5. For example, $2x^2 + x + 3 + x^2 - 3x$
6. a) $7x - 2$ b) $-2x^2 - x$
   c) $5x^2 - 2$ d) $3x + 11$
   e) $7x^2 + 3x - 2$ f) $3x^2 + 4x - 1$
   g) $4x^2 - 7x + 1$
7. Use algebra tiles.
   a) $9x - 3$ b) $-x^2 + 4x$
   c) $-2x^2 - 7x + 12$ d) $3x^2 - 3x + 11$
8. a) -12 b) $-2x^2 + 9$
   c) $2x^2 - 5x$ d) $2x^2 + 2x + 2$
   e) $-8x^2 + 6x + 5$ f) $x^2 - 4x + 6$
9. a) $15x^2 + 20$ b) $28 + 8x$
   c) $3x^3 - 6x^2 + 6$ d) $16x^2 + 24x - 32$
   e) $14x^3 - 7$
10. a) $-6x - 12$ b) $-7x^2 + 28$
    c) $-4x^2 - 2$ d) $-15x^3 + 10x + 15$
    e) $-3x^3 + 6x^2 - 12$
11. $16x - 4; 20x - 4; 24x - 4$
12. a) b)
13. a) i) $24x^2$ ii) $12x^3$ iii) $-12x^2$ iv) $18x^3$
    b) Parts i and iii. The other parts would require cubes.
14. a) $3x^2 + 4x$ b) $3x^3 - 24x$
    c) $8x - 2x^2$ d) $-18x^3 + 24x^2 + 12x$
    You can use algebra tiles for parts a, b, and c. You cannot use them for part d because no tile represents x^3.
15. a) $6x^2 - 9x$ b) $-2x^3 + 10x$
    c) $-15x^2 - 35x$ d) $8x^3 + 10x^2 - 6x$
16. a) $12x^3; 14x^3; 16x^3$
    b) $4x^2 - 12x; 5x^2 - 15x; 6x^2 - 18x$
17. a) $6x + 1 = 2x + 13$ b) $x = 3$ c) 3
18. a) $x = 2$ b) $x = 2$ c) $x = -3$
19. a) $x = 6$ b) $x = -4$
    c) $x = -1$ d) $x = 3$
20. a) $x = 1000$
    b) The number of tickets that must be sold to raise $60 000 after expenses.

Chapter 7 Practice Test, page 290

1. C
2. C
3. a) $5x^2 - 4x - 5$ b) $-x^2 + 2x$
   c) $3x + 12$ d) $6x^3$ e) $4x^3 - 20x^2 + 12x$
4. a) $4900 b) 320 students
5. A polynomial can be simplified if it contains like terms. For example, $2x + 4 + 3x$ simplifies to $5x + 4$.
6. a) Joe forgot to subtract every term of the second polynomial.
   b) $2x^2 + 2x - 4$
   c) Add the difference to the second polynomial.

Cumulative Review Chapters 1–7, page 291

1. a) $P \doteq 29.7$ cm; $A \doteq 51$ cm^2
   b) $P \doteq 31.4$ cm; $A \doteq 33.9$ cm^2
2. a) About 62.8 cm^3 b) About 23.9 cm^3
   c) 1440 cm^3 d) 400 cm^3
   e) About 7238.2 cm^3
3. a) i) 10 cm by 10 cm
      ii) 17 m by 17 m
      iii) 22.5 cm by 22.5 cm
   b) i) 100 cm^2 ii) 289 m^2
      iii) 506.25 cm^2
      For a given perimeter, the rectangle with the greatest area is a square.
4. 12 cm by 12 cm; 48 cm
5. 20 units

 or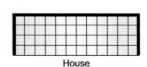

6. No; the sum of the angles in a triangle is 180°. The largest angle in a right triangle is 90°.
7. a) $a = 75°, b = 75°, c = 105°, d = 150°$
   b) $y = b = d = 100°; a = c = x = z = 80°$
   c) $m = 90°, n = p = 110°$
8. a) 72° b) 40° c) 36°
9. a) $n = 15$ b) $b = 18$
   c) $a = 6$ d) $m = 20$

10. a) 72 km/h b) About $1.26/kg
    c) $8.50/h d) 3.6 km/h
    e) About $0.58/muffin
11. a) 12 lengths b) 22.5 min
12. a) $24.49 b) $28.16
13. a) $37.50 b) $2537.50
14. a) World records for women's discus throw
    b) About 70.2 m, about 71.5 m
    c) 1974
    d)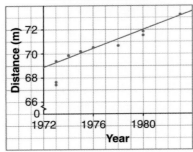
    e) Extend the graph to the right: About 73.5 m
15. a) Yes, as the width increases by 1 cm, the length decreases by 1 cm.
    b) Yes, the graph is a straight line.
    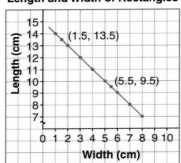
    c) i) 9.5 cm ii) 1.5 cm
    d) Length plus width is 15 cm.
16. a), b)
    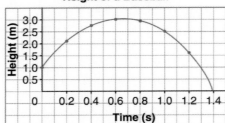
    c) About 3.0 m d) About 1.39 s
    e) At about 0.18 s and at about 1.12 s
17. a) From A to B, Tyler drives at an average speed of 60 km/h for 30 min and travels 30 km.
       From B to C, Tyler stops driving for about 15 min.
       From C to D, Tyler drives at an average speed of 80 km/h for 45 minutes, and travels 60 km.
       From D to E, Tyler stops driving for 5 min.
       From E to F, Tyler drives at an average speed of 60 km/h for 40 min and travels 40 km.
    b) From B to C, Tyler buys gas (it is a longer stop). From D to E, Tyler picks up his cousin (it is a shorter stop).
18. a) Frame 4:
    b) The relation is linear because the first differences are equal.

Frame number	Number of toothpicks	First differences
1	3	
		2
2	5	
		2
3	7	
		2
4	9	
		2
5	11	
		2
6	13	

    c) Yes, the points lie on a straight line.
    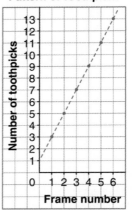
19. a) Yes, the graph is a straight line that passes through the origin.
    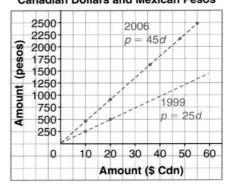

b) 45 pesos/$; for $1, you can get 45 pesos.
c) $p = 45d$
d) 7875 pesos
e) About $166.67
f) Answers may vary.

20. a) The new graph would be less sharp. (Refer to graph in question 19 part a).
b) $p = 25d$

21. a)

Time (h)	Cost ($)
0	50
10	85
20	120
30	155
40	190
50	225
60	260

b) No, the graph does not pass through the origin.

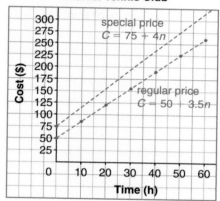

c) $C = 50 + 3.5n$ d) About 42 h

22. a) The vertical intercept is greater and the graph is steeper. (Refer to graph in question 21 part b).
b) $C = 75 + 4n$ c) About 31 h

23. a) $C = 40 + n$ b) $R = 3n$
c)

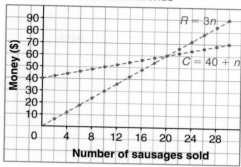

d) 20 sausages; then revenue is equal to cost.

24. a) $x^2 + 6x - 5$
b) $-2x^2 - 3x + 2$

25. a) $7x + 3$
b) $2x^2 - 3x$
c) 5
d) $5x^3 + 4x$

26. a) $11x - 4$
b) $-x^2 - 5x$
c) $2x^2 - 4x + 1$

27. a) $x + 3$
b) $x^2 - 5$
c) $7x^2 - 2x + 1$

28. a) $30 + 12x$
b) $2x^3 - 6x^2 + 6$
c) $-9x - 3$
d) $-8x^3 + 20x^2$
e) $2x^2 + 7x$
f) $21x - 9x^2$
g) $-4x^3 + 6x$
h) $6x^3 - 8x^2 + 10x$

29. a) $x = 1$
b) $x = -3$
c) $x = -3$
d) $x = 2$

Acknowledgments

Pearson Education would like to thank the Royal Canadian Mint for the illustrative use of Canadian coins in this textbook. In addition, the publisher wishes to thank the following sources for photographs, illustrations, and other materials used in this book. Care has been taken to determine and locate ownership of copyright material in this text. We will gladly receive information enabling us to rectify any errors or omissions in credits.

Photography

Cover: Carlos Dominguez / Science Photo Library; p. 1 Royalty-Free/CORBIS; p. 2 Michael S. Yamashita/CORBIS; p. 3 (top) Photonica/Philip J. Brittan, p. 3 (bottom) Dave Starrett; p. 7 Bill Miles/CORBIS; p. 11 Veer Incorporated/Beth Dixson; p. 15 Ryan McVay/Photodisc/Getty Images; p. 18 Dave Starrett; p. 19 Dave Starrett; p. 21 (top) Stockbyte; p. 21 (bottom) Ray Boudreau; p. 25 (top) Digital Vision/Getty Images; p. 25 (bottom) Dave Starrett; p. 29 (top) Royalty-Free/CORBIS; p. 29 (bottom) Dave Starrett; p. 30 Klaus Leidorf/zefa/CORBIS; p. 31 Ray Boudreau; p. 33 (top) Wolfram Schroll/zefa/CORBIS; p. 33 (bottom left and right) Dave Starrett; p. 41 image100/MaXx Images; p. 43 Ian Crysler; p. 48 Ian Crysler; p. 49 Ian Crysler; p. 53 (top) Spencer Grant/PhotoEdit Inc.; p. 59 Ian Crysler; p. 64 Ian Crysler; p. 65 Ray Boudreau; p. 73 Hisham F Ibrahim/Photodisc/Getty Images; p. 75 Ian Crysler; p. 81 (top) Plantography/Alamy; p. 81 (bottom) Ian Crysler; p. 88 Ian Crysler; p. 91 Ian Crysler; p. 95 (top) Ian Crysler; p. 101 David Cooper/Toronto Star/firstlight.com; p. 102 Royalty-Free/CORBIS; p. 109 Pixtal/MaXx Images; p. 111 (top) Michael Newman/PhotoEdit Inc.; p. 111 (bottom) Ian Crysler; p. 115 Ian Crysler; p. 116 photos.com Images/Jupiter Images Unlimited; p. 117 Ian Crysler; p. 118 Ian Crysler; p. 121 (top) Comstock.com Images/Jupiter Images Unlimited; p. 121 (bottom) Dave Starrett; p. 122 (left) Wally Bauman/Alamy; p. 122 (right) Mika/Zefa/CORBIS; p. 127 (top) Yellow Dog Productions/The Image Bank/Getty Images; p. 130 Zaw Min Yu/Lonely Planet Images; p. 131 Scott Cunningham/Getty Images; p. 135 (top left) Frank Siteman/PhotoEdit Inc.; p. 135 (top centre) The Garden Picture Library/Alamy; p. 135 (top right) Dennis MacDonald/PhotoEdit Inc.; p. 135 (bottom) Dave Starrett; p. 145 Brand X Images/Jupiter Images Unlimited; p. 147 (top) Duomo/CORBIS; p. 147 (bottom) Ray Boudreau; p. 148 Yann Arthus-Bertrand/CORBIS; p. 151 (top) Stewart Cohen/Pam Ostrow/Blend Images/Getty Images; p. 151 (centre and bottom) Ian Crysler; p. 154 Ian Crysler; p. 156 Comstock.com Images/Jupiter Images Unlimited; p. 159 Ian Crysler; p. 160 David Parker/Photo Researchers, Inc.; p. 164 Ian Crysler; p. 165 Ian Crysler; p. 167 Ian Crysler; p. 169 (top) Ian Crysler; p. 169 (bottom) Ray Boudreau; p. 171 Ian Crysler; p. 173 Ray Boudreau; p. 175 Dorling Kindersley; p. 176 Georgette Douwma/Photographer's Choice/Getty Images; p. 181 (top) Ian Crysler; p. 181 (bottom) Dave Starrett; p. 189 Comstock.com Images/Jupiter Images Unlimited; p. 191 Ian Crysler; p. 193 Dave Starrett; p. 197 Ian Crysler; p. 200 Ian Crysler; p. 201 Tom Stewart/CORBIS; p. 202 Bo Zaunders/CORBIS; p. 205 (top) Steve Allen/Brand X Images/Alamy Images; p. 210 Ian Crysler; p. 211 Courtesy of the Green Party of Canada; p. 214 David Else/Lonely Planet Images; p. 217 Adrian Wyld/Canadian Press; p. 219 Ian Crysler; p. 221 Ian Crysler; p. 227 Ian Crysler; p. 230 Frank Whitney/Ionica/Getty Images; p. 231 Gregory K. Scott/Photo Researchers, Inc.; p. 233 Hemera/MaXx Images; p. 235 Canadian Press NATARK; p. 241 (top) Paul Chiasson/Canadian Press; p. 241 (bottom) Ian Barrett/Canadian Press; p. 245 Photodisc/Getty Images; p. 246 Jeff Greenberg/Index Stock Imagery; p. 251 David Young-Wolff/PhotoEdit Inc.; p. 257 Ian Crysler; p. 260 Ian Crysler; p. 263 Ian Crysler; p. 267 Ian Crysler; p. 275 (top) Ian Crysler; p. 279 Ian Crysler; p. 283 David Frazier/PhotoEdit Inc.; p. 290 Spencer Grant/PhotoEdit Inc.

Illustrations

Steve Attoe, Philippe Germain, Brian Hughes, Allan Moon, Neil Stewart, Joe Weissmann, Rose Zgodzinski

p. 74 From Frayer, D.A., Frederick, W.C. & Klausmeier, H.J. (1969). *A scheme for testing the level of concept mastery* (Working Paper No. 16). Madison: Wisconsin Research and Development Center for Cognitive Learning. Reprinted with permission.

Fathom Dynamic Statistics, Key Curriculum Press, 1150 65th Street, Emeryville, CA 94608, 1-800-995-MATH, www.keypress.com/fathom.
The Geometer's Sketchpad, Key Curriculum Press, 1150 65th Street, Emeryville, CA 94608, 1-800-995-MATH, www.keypress.com/sketchpad.
Texas Instruments images used with permission.